Understanding the Connections Between

COASTAL WATERS *and* OCEAN ECOSYSTEM SERVICES *and* HUMAN HEALTH

Workshop Summary

Rose Marie Martinez and Erin Rusch, *Rapporteurs*

Roundtable on Environmental Health Sciences, Research, and Medicine

Board on Population Health and Public Health Practice

INSTITUTE OF MEDICINE
OF THE NATIONAL ACADEMIES

THE NATIONAL ACADEMIES PRESS
Washington, D.C.
www.nap.edu

THE NATIONAL ACADEMIES PRESS 500 Fifth Street, NW Washington, DC 20001

NOTICE: The workshop that is the subject of this workshop summary was approved by the Governing Board of the National Research Council, whose members are drawn from the councils of the National Academy of Sciences, the National Academy of Engineering, and the Institute of Medicine.

This activity was supported by contracts between the National Academy of Sciences and the National Institute of Environmental Health Sciences (HHSN26300033), The Kresge Foundation, Colgate-Palmolive Company, ExxonMobil Foundation, and Royal Dutch Shell. The views presented in this publication do not necessarily reflect the views of the organizations or agencies that provided support for the activity.

This summary is based on the proceedings of a workshop that was sponsored by the Roundtable on Environmental Health Sciences, Research, and Medicine. It is prepared in the form of a workshop summary by and in the name of the rapporteurs as an individually authored document.

International Standard Book Number-13: 978-0-309-29468-3
International Standard Book Number-10: 0-309-29468-1

Additional copies of this report are available from the National Academies Press, 500 Fifth Street, NW, Keck 360, Washington, DC 20001; (800) 624-6242 or (202) 334-3313; http://www.nap.edu.

For more information about the Institute of Medicine, visit the IOM home page at: **www.iom.edu.**

Printed in the United States of America

The serpent has been a symbol of long life, healing, and knowledge among almost all cultures and religions since the beginning of recorded history. The serpent adopted as a logotype by the Institute of Medicine is a relief carving from ancient Greece, now held by the Staatliche Museen in Berlin.

Suggested citation: IOM (Institute of Medicine). 2014. *Understanding the connections between coastal waters and ocean ecosystem services and human health: Workshop summary.* Washington, DC: The National Academies Press.

"Knowing is not enough; we must apply.
Willing is not enough; we must do."
—Goethe

INSTITUTE OF MEDICINE
OF THE NATIONAL ACADEMIES

Advising the Nation. Improving Health.

THE NATIONAL ACADEMIES

Advisers to the Nation on Science, Engineering, and Medicine

The **National Academy of Sciences** is a private, nonprofit, self-perpetuating society of distinguished scholars engaged in scientific and engineering research, dedicated to the furtherance of science and technology and to their use for the general welfare. Upon the authority of the charter granted to it by the Congress in 1863, the Academy has a mandate that requires it to advise the federal government on scientific and technical matters. Dr. Ralph J. Cicerone is president of the National Academy of Sciences.

The **National Academy of Engineering** was established in 1964, under the charter of the National Academy of Sciences, as a parallel organization of outstanding engineers. It is autonomous in its administration and in the selection of its members, sharing with the National Academy of Sciences the responsibility for advising the federal government. The National Academy of Engineering also sponsors engineering programs aimed at meeting national needs, encourages education and research, and recognizes the superior achievements of engineers. Dr. C. D. Mote, Jr., is president of the National Academy of Engineering.

The **Institute of Medicine** was established in 1970 by the National Academy of Sciences to secure the services of eminent members of appropriate professions in the examination of policy matters pertaining to the health of the public. The Institute acts under the responsibility given to the National Academy of Sciences by its congressional charter to be an adviser to the federal government and, upon its own initiative, to identify issues of medical care, research, and education. Dr. Harvey V. Fineberg is president of the Institute of Medicine.

The **National Research Council** was organized by the National Academy of Sciences in 1916 to associate the broad community of science and technology with the Academy's purposes of furthering knowledge and advising the federal government. Functioning in accordance with general policies determined by the Academy, the Council has become the principal operating agency of both the National Academy of Sciences and the National Academy of Engineering in providing services to the government, the public, and the scientific and engineering communities. The Council is administered jointly by both Academies and the Institute of Medicine. Dr. Ralph J. Cicerone and Dr. C. D. Mote, Jr., are chair and vice chair, respectively, of the National Research Council.

www.national-academies.org

PLANNING COMMITTEE FOR THE WORKSHOP ON UNDERSTANDING THE CONNECTIONS BETWEEN COASTAL WATERS AND OCEAN ECOSYSTEM SERVICES AND HUMAN HEALTH[1]

JAY LEMERY, University of Colorado, Denver, CO
PAUL SANDIFER, National Oceanic and Atmospheric Administration, Charleston, SC
CARL SHAPIRO, U.S. Geological Survey, Reston, VA
JOHN SPENGLER, Harvard University, Boston, MA
HAROLD ZENICK, U.S. Environmental Protection Agency, Washington, DC

[1] Institute of Medicine planning committees are solely responsible for organizing the workshop, identifying topics, and choosing speakers. The responsibility for the published workshop summary rests with the workshop rapporteurs and the institution.

ROUNDTABLE ON ENVIRONMENTAL HEALTH SCIENCES, RESEARCH, AND MEDICINE[1]

[1] Institute of Medicine forums and roundtables do not issue, review, or approve individual documents. The responsibility for the published workshop summary rests with the workshop rapporteurs and the institution.

Reviewers

This workshop summary has been reviewed in draft form by individuals chosen for their diverse perspectives and technical expertise, in accordance with procedures approved by the National Research Council's Report Review Committee. The purpose of this independent review is to provide candid and critical comments that will assist the institution in making its published workshop summary as sound as possible and to ensure that the workshop summary meets institutional standards for objectivity, evidence, and responsiveness to the study charge. The review comments and draft manuscript remain confidential to protect the integrity of the process. We wish to thank the following individuals for their review of this workshop summary:

Samantha Joye, University of Georgia
Kimberley Thigpen-Tart, National Institute of Environmental Health Sciences
Juli Trtanj, National Oceanic and Atmospheric Administration

Although the reviewers listed above have provided many constructive comments and suggestions, they did not see the final draft of the workshop summary before its release. The review of this workshop summary was overseen by **Richard B. Johnston,** University of Colorado School of Medicine. Appointed by the Institute of Medicine, he was responsible for making certain that an independent examination of this workshop summary was carried out in accordance with institutional procedures and that all review comments were carefully considered. Responsibility for the final content of this workshop summary rests entirely with the rapporteurs and the institution.

Contents

1

Introduction

AIMS OF THE WORKSHOP

Lynn R. Goldman, M.D., M.P.H.
Vice Chair, Roundtable on Environmental Health Sciences, Research, and Medicine
Dean, George Washington University School of Public Health

Lynn Goldman set the stage for the workshop by noting that the Roundtable on Environmental Health Sciences, Research, and Medicine was established in 1998 to allow science and technology leaders in environmental health, health care organizations, top administrators in government agencies, representatives from consumer interest groups, representatives in industry, and academicians to convene and discuss current and emerging environmental health issues. The Roundtable seeks not to reach consensus on any particular issue, but rather to engage in rigorous dialogue and sharing of perspectives to forge new directions and solutions. She noted that the definition of the environment adopted by the Roundtable is a broad one—it includes the natural, built, and social environments and their impact on human health. In conducting its work, the Roundtable considers issues from a life-course perspective (i.e., how the environment affects individuals from birth to death) and a life-cycle perspective (i.e., beginning to the end of a process), and it uses multidisciplinary approaches to assess the impacts of environmental change on human health. She provided examples of workshop topics the Roundtable has considered: urban sprawl and its impact on human health, the role of nanotechnology and environmental health, and, most recently, shale gas extraction and human health. These workshop discussions help to illuminate and build understanding of the complex environmental factors that impact human health and provide information that scientists and decision makers can use to promote healthier environments.

The Roundtable chose to host a workshop to examine the relationship between ecosystem services and human health, and in doing so is forging a new direction for environmental public health, Goldman said. Ecosystem services are the benefits that people obtain from ecosystems. For example, humans rely on the natural environment (e.g., oceans and coastal waters) for essential human services such as providing food and essential dietary nutrients and purifying drinking water. Humans also rely on the environment for the natural cycles that renew the oxygen in the air, reduce carbon dioxide in the air, and recycle nitrogen, Goldman said. The Millennium Ecosystem Assessment (MEA, 2005) concluded that approximately 60 percent of 24 ecosystem services were being degraded or used unsustainably—this percentage is not likely to be lower today, she stated. She noted that while ecosystem services are interdependent, decision makers manage each of these services independently of each other. She concluded that the result can be devastating. For the coastal areas, where a majority of the global populations are living, the loss of mangroves and barrier islands has resulted in significant loss of human health due to coastal flooding. Droughts in the Midwest have resulted in saltwater intrusion of the Mississippi River. For New Orleans parishes, this change requires moving their drinking water intake pipes farther upstream. Goldman noted that these scenarios will continue as climate change continues to be unabated and the economic costs will be staggering.

With growing populations and increasing standards of living globally, additional pressures will be on the world's oceans and waterways to provide essential services. During the workshop, Goldman noted there is a need to discuss the current and future availability and demand for seafood, which currently shows no sign of subsiding. In fact, the Institute of Medicine's (IOM's) report *Seafood Choices: Balancing Benefits and Risks* (IOM, 2006) highlighted the health benefits of the consumption of seafood that is relatively free of contaminants. Fisheries and fish farms will need to be carefully managed for all people to have access to this vital source of omega-3 fatty acids and protein. Currently, many regions around the United States and around the world are in danger of losing these crucial services, she said.

Goldman highlighted one example of health effects in an area that lost ecosystem services. The loss of the Aral Sea, due to the diversion of water for irrigation, has been accompanied by many worsening health indices, including increases in infant mortality in the region, rising from 25 per 1,000 live births in 1950 to 70 to 100 per 1,000 live births in

1996. In comparison, she noted that the U.S. infant mortality rate was approximately 7 per 1,000 in 1996. At the same time, the levels of persistent toxics such as organochlorine pesticides, polychlorinated biphenyls, and dioxins have been found at very high levels in blood and breast milk in the Aral Sea region as a result of the overexploitation of the region for growing cotton and other crops. Whether there is a causal association between the levels of toxics and declining health in the Aral Sea region is debatable; however, it is clear, Goldman noted, that without a new approach to ecosystem services, and especially those provided by our multiple uses of ocean and coastal ecosystems, the health of individuals in the United States could be similarly threatened.

This workshop provided an opportunity to discuss coastal waters and ocean ecosystem services in the United States and to understand impacts on human health and elucidate key linkages by discussing three questions:

1. What ecosystem services provided by coastal waterways and oceans are essential for human health and well-being? How can these services be valued and what are the consequences of these services being reduced or not provided?
2. What major stressors, both natural and human induced, can affect the ability of coastal waterways and ocean systems to provide essential services?
3. How can we best consider these impacts in decision making? What are the key factors that affect the resiliency of these systems and how can we enhance their resilience?

STRUCTURE OF THE SUMMARY

The workshop was organized by an independent planning committee,[1] whose role was limited to planning the workshop, in accordance with the procedures of the National Research Council. This summary was prepared by the workshop rapporteurs as a factual summary of what

---

[1] The planning committee's role was limited to planning the workshop, and the workshop summary has been prepared by the workshop rapporteurs as a factual summary of what occurred at the workshop. Statements, recommendations, and opinions expressed are those of individual presenters and participants and are not necessarily endorsed or verified by the Institute of Medicine, and they should not be construed as reflecting any group consensus.

occurred at the workshop. All views presented in the summary are those of the workshop participants. The summary does not contain any findings or recommendations by the planning committee or the Roundtable. The statement of task of the workshop can be found in Box 1-1.

The presentations and discussions that occurred during the workshop are summarized in the subsequent chapters. Chapter 2 provides an overview of the conceptual issues surrounding coastal waters and ocean ecosystem services and environmental health. Chapter 3 covers stressors impacting coastal and ocean ecosystem services and human health. Chapter 4 explores connections between seafood supplies and food availability. Chapter 5 considers opportunities for medicines arising from ecosystem services. Chapter 6 follows with links between coastal change and human health. Chapter 7 examines how the benefits of recreational waters can be maintained through monitoring. Chapter 8 presents new approaches for optimizing sustainable solutions to protect ecosystem services and human health. Chapter 9 provides a brief summation of key points discussed at the workshop. The workshop agenda can be found in Appendix A and the workshop speaker biosketches are included in Appendix B.

BOX 1-1
Statement of Task

An ad hoc committee will plan and conduct a public workshop on the connection of ecosystem services and human health. The committee will focus on oceans, but may draw from other areas of the natural environment. The workshop will feature invited presentations and discussions to look at the state of the science of the role of oceans in ensuring human health and identify gaps and opportunities for future research. The committee will develop the workshop agenda, select invited speakers and discussants, and moderate the discussions. A workshop summary will be prepared by a designated rapporteur in accordance with National Research Council policies and procedures.

REFERENCES

IOM (Institute of Medicine). 2006. *Seafood choices: Balancing benefits and risks*. Washington, DC: The National Academies Press.

MEA (Millennium Ecosystem Assessment). 2005. *Millennium ecosystem assessment. Synthesis report*. Washington, DC: Island Press.

2

Conceptual Issues

This chapter provides a summary of presentations outlining the conceptual issues of coastal waters and ocean ecosystem services and human health. The first presentation gives an overview of the concept of ecosystem services and ways to use this in natural resource management decision making and other practices of corporations and governments. The second presentation describes broad efforts to protect ecosystem services, using the Puget Sound and Gulf of Mexico as examples, and the need to develop indicators to assess the health of coastal and ocean environments and the services they provide to humans. The presentations are followed by a summary of the discussion that ensued.

UNDERSTANDING ECOSYSTEM SERVICES

Lydia Olander, Ph.D.
Director, Ecosystem Services Program,
Nicholas Institute, Duke University

Lydia Olander began the presentation by defining the concept of ecosystem services as the benefits people obtain from ecosystems.[1] The term *ecosystem services* is not an entirely new concept, she said. The notion of ecosystem services was included in the thinking of Gifford Pinchot, the first chief of the Forest Service, in the 1900s when he discussed conservation as the wise use of the earth and its resources for the lasting good of men. In contrast to the past, today's concept of

---

[1] "An ecosystem is a dynamic complex of plant, animal, and microorganism communities and the nonliving environment interacting as a functional unit." Humans are an integral part of ecosystems (MEA, 2005).

ecosystem services goes beyond the typical services of timber production, grazing lands, crop production, and fish production to include the categorization, quantification, and valuation of a broader range of services. Additionally, she noted, the field is grappling with how to operationalize the concept for use in resource management decision making, and how to incorporate it into standard practices. The concept is moving beyond the realm of ecosystem scientists and economists to the decisions and actions of conservation organizations, governments, and corporations.

According to Olander, one of the key events that changed thinking in the scientific community in the early 1990s was Biosphere 2, a research effort to recreate the biosphere and have a self-sustaining environment. Biosphere 2's initial experiments failed to sustain human life in a closed and independent biosphere system as hoped. The biosphere experienced a broad range of negative impacts: oxygen levels plummeted, carbon dioxide levels skyrocketed, nitrous oxide rose to levels that impair brain function, and water purification failed. Further, of 25 vertebrate species studied, 19 died. Pollinators became extinct, and ants, cockroaches, and katydids became rampant. These results highlighted the fact that ecosystem services, the underpinning of human well-being, are not well understood and are difficult to recreate.

Millennium Ecosystem Assessment and Ecosystem Changes

Olander proceeded to describe the Millennium Ecosystem Assessment (MEA) (2005), an extensive effort to provide an integrated assessment of the consequences of ecosystem change for human well-being. The MEA identifies four categories of ecosystem services that lead to health and well-being: (1) provisioning, (2) regulating, (3) cultural, and (4) supporting services.

Provisioning services are the typical goods produced or provided by ecosystems. These include foods such as crops, livestock, fish from capture fisheries and aquaculture, and wild foods. Fibers are another product category and typically include timber, cotton, hemp, silk, and wood fuel. In addition, genetic resources, freshwater provisions, biochemicals, natural medicines, and pharmaceuticals obtained from the environment are ecosystem provisions. These provisions tend to have a market value.

Regulation and sustainability of ecosystem processes is another type of service and an important contributor to human health. These processes include climate regulation, air quality regulation, erosion regulation, water purification, disease and pest regulation, pollination, and natural hazard regulation such as flood protection.

The category of cultural services is more abstract and includes non-material benefits obtained from ecosystems. Services such as spiritual and religious values, knowledge systems, education values, inspiration, aesthetic values, social relations, sense of place and recreation, ecotourism, and others are among those included in this category. Olander described an example of another cultural service that is not typically considered—darkness. In Puerto Rico, darkness is a very important part of the ecosystem and is critical to ecotourism and recreation. Bioluminescent bays, for example, draw many ecotourists but light pollution is a problem.

The category of supporting services includes those services necessary for the production of all other ecosystem services such as primary production, production of oxygen or photosynthesis, and soil formation. Olander highlighted that biodiversity is underpinning all services.

The MEA framework has been used in two ways, Olander said. The first way is in tracking the status of the environment as it relates to people or assessing trends in ecosystem services. The second way is in improving the management of natural resources. For example, the MEA framework was used to understand increased flooding across the globe. The MEA can help illustrate how flooding affects people; the impact can be seen in deaths, use of health care services, disease vectors, and many others. But it is useful in identifying and understanding the causes or drivers of flooding. Two reasons were identified as contributing to the human impact of flooding: (1) people increasingly occupy regions exposed to extreme events and (2) significant losses of coastal habitat reduces flood protection and increases the risk of floods. The latter describes the loss of ecosystem services. Loss of coastal habitat can be tracked as an indicator of increasing flood risk. In addition, by focusing on this driver, linkages to other services that might be lost or gained can be examined to understand trade-offs and synergies. The loss of coastal habitat can reduce shellfish and fish nurseries, which can result in fewer fish, fewer fish-related jobs, higher costs of fish, and fewer fresh fish available for consumption (the latter could have an impact on health). Further, the loss of coastal habitat can lead to a reduction in nitrogen transformation and storage which can increase the impact of nitrogen loading and increase the risk of eutrophication (high nutrient levels) leading to coastal dead zones and fish loss. But there can also be positive impacts such as better wave energy because sea grasses are not in the way (e.g., to generate energy from surface waves). Both the negative and positive impacts must be considered.

The primary finding of the MEA is that 60 percent (15 out of 24) of the ecosystem services examined are being degraded or used unsustainably (MEA, 2005). Ecosystem services that have been degraded include capture fisheries, water supply, waste treatment and detoxification, water purification, natural hazard protection, regulation of air quality, regulation of regional and local climate, regulation of erosion, spiritual fulfillment, and aesthetic enjoyment. The use of two ecosystem services in particular—capture fisheries and freshwater—are now well beyond levels that can be sustained at current demands, much less future levels, Olander said.

An aspect of the MEA that needs to be highlighted, Olander noted, is the effort to build linkages between the services and human well-being. As can be seen in Figure 2-1, ecosystem services are linked to constituents of well-being—security, basic material for a good life, health, good social relations, and freedom of choice and actions. The arrow colors identify the potential for mediation by socioeconomic factors, and the arrow widths identify the intensity of linkages between ecosystem services and human well-being.

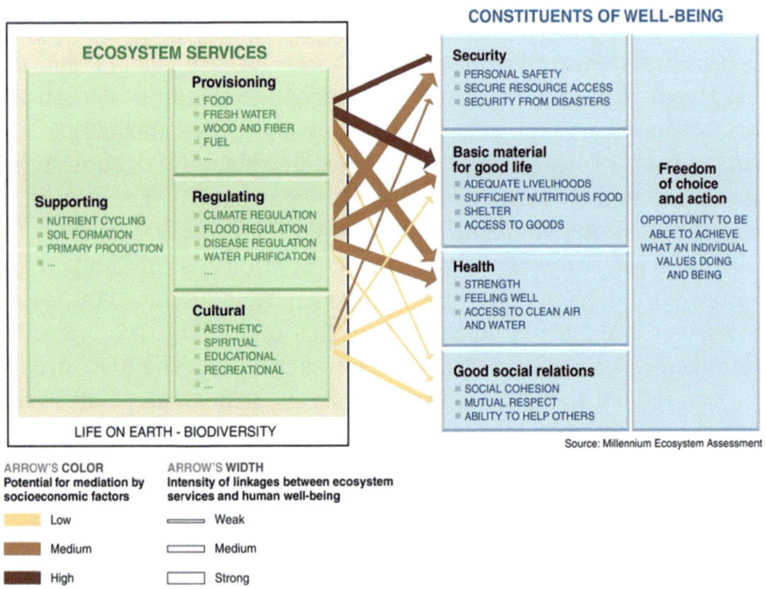

FIGURE 2-1 Linkages between ecosystem services and human well-being. SOURCE: MEA, 2005. Reprinted with permission from the Millennium Ecosystem Assessment.

Ecosystem Services Framework to Improve Natural Resource Management Decisions

Olander continued with a discussion of the use of an ecosystems services framework to improve the management of natural resources. It is also important to understand how changes in resource management decisions affect the production of ecosystem services, she said. This relationship between management and actions and ecosystem services outcomes is called a production function by the scientific community. To further explain a production function, consider that X percent of habitat loss results in a Y percent decrease in coastal flood protection provided by the ecosystem, which results in a Z percent increase in an impact (for example, mortality). The question then is, how much reduction in coastal habitat, of which types and where, result in how much of an increase in flooding and of what type and where?

Olander noted that there are trade-offs and synergies in the management of natural resources. A small change in coastal habitat may have a large impact on fisheries but only a small impact on flood protection. So it is important to understand the relationships across services. Further, if decisions need to be made, the trade-offs need to be evaluated. For example, people want to live in coastal zones for recreational, spiritual, and economic opportunities. But housing development and infrastructure can damage the services the coastal systems produce. Ideally ecosystem services approaches can make the trade-offs inherent in management choices (across services that matter to different people) explicit.

Olander described a compelling example from the MEA related to mangroves. Natural mangroves provide many services, including nursery and adult fishery habitat, fuel wood and timber, and carbon sequestration; they detoxify pollutants and provide protection from erosion and natural disasters. However, the promotion of shrimp farming in these areas has raised concerns about the potential ecological and economic impact of clearing mangroves for shrimp farming. Researchers studying the value of mangrove conservation showed that while financial cost-benefit analysis demonstrates that it is financially worthwhile for an individual to convert mangrove forests into commercial shrimp farms, once other factors are taken into account, such as the water pollution from shrimp ponds and restoring the mangrove after the shrimp pond is abandoned, shrimp farming is no longer as economically beneficial to society (Sathirathai and Barbier, 2001). The study demonstrated that consideration of other values and changes in services may lead to different decisions.

Olander noted that there are a number of ways to approach valuation and comparing alternatives. Some are qualitative in nature, for example an ecosystem services management effects matrix (see Figure 2-2). The matrix allows decision makers to consider a range of management options and to assess their impact on ecosystem services against a gradient from strong positive to strong negative impact.

Olander also described another framework commonly used by researchers, published by de Groot and colleagues (2010), which is useful for linking ecosystem services to human well-being in a way that facilitates quantification of impact. As shown in Figure 2-3, ecosystem services are generated by ecosystem functions. Ecosystem functions are intermediate between ecosystem processes and services and are generally described as the capacity of the ecosystem to provide goods and services that can benefit people. The actual use of the service provides benefits, and those benefits can be valued.

Olander noted that value can be assessed based on stakeholder preferences or economic costs and benefits; however, economic value may be difficult to impossible to assess for some services. For example, market values are easier to assess for food services which are valued by the market than for cultural services such spiritual and religious benefits. A range of qualitative to quantitative ecosystem tools are available to value services. These tools continue to be evaluated and new tools are being developed. Below are some examples that show the range of different types of tools available:

- Ecosystem Services Review is a structured methodology for corporate managers to proactively develop strategies for managing risks and opportunities arising from their company's dependence and impact on ecosystems (Hanson et al., 2012).
- Integrated Valuation of Ecosystem Services and Tradeoff (InVEST)[2] models are based on production functions that define how an ecosystem's structure and function affect the flows and values of ecosystems (Tallis et al., 2010).
- Social Values for Ecosystem Services (SolVES)[3] is a geographic information system (GIS) application for assessing, mapping, and

[2] InVEST is available at: http://www.naturalcapitalproject.org/download.html (accessed August 25, 2013).

[3] SolVES is available at: http://solves.cr.usgs.gov (accessed August 25, 2013).

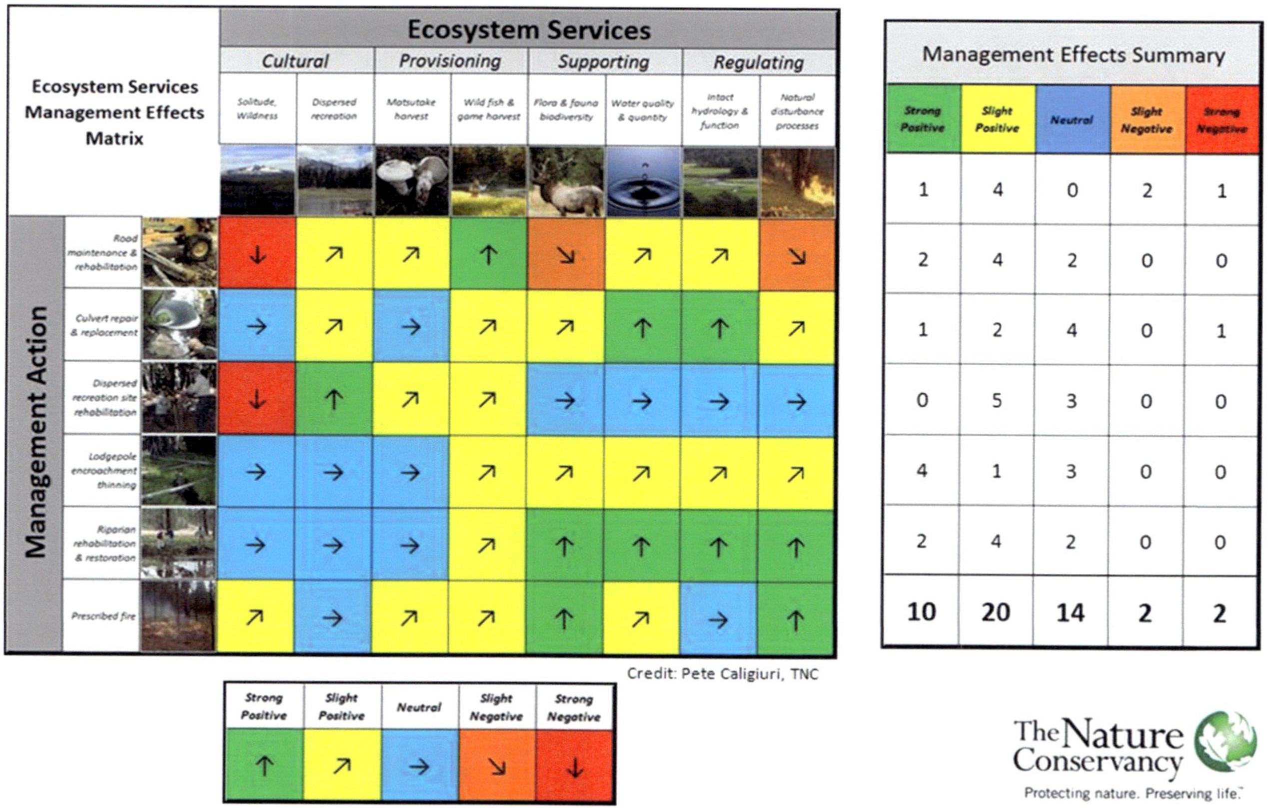

FIGURE 2-2 Ecosytem services management effects matrix applied to a marsh scenario.
SOURCE: Personal communication, Pete Caligiuri, The Nature Conservancy, September 11, 2013. Reprinted with permission.

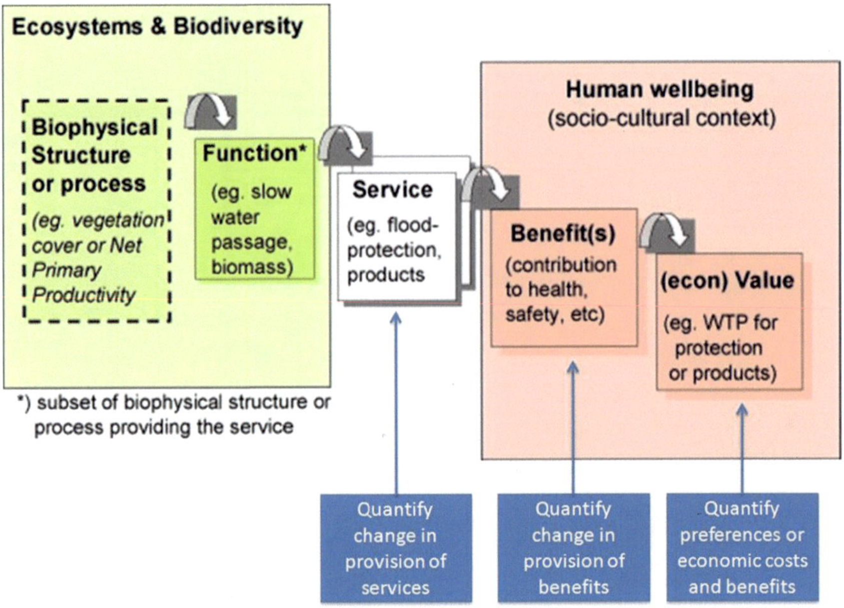

FIGURE 2-3 Framework for linking ecosystems and biodiversity to human well-being.
NOTE: econ = economic, WTP = willingness to pay.
SOURCE: Adapted from de Groot et al., 2010. Reprinted with permission from Elsevier.

quantifying the social values of ecosystem services (Sherrouse et al., 2011).

- Artificial Intelligence for Ecosystem Services (ARIES)[4] is a Web-based technology to assist rapid ecosystem service assessment and valuation (Bagstad et al., 2011).

In summary, Olander highlighted that the concept of ecosystem services is a logical thought process which can be used in quantitative or qualitative assessment of implications and trade-offs. Efforts are now focused on figuring out how to implement these methods more broadly and to develop decision-making tools, which continues to be a complex challenge.

---

[4] ARIES is available at: http://www.ariesonline.org (accessed August 25, 2013).

INTEGRATION OF ENVIRONMENTAL HEALTH AND MARINE ECOSYSTEM SERVICES

Tracy Collier, Ph.D.
Science Director, Puget Sound Partnership

Tracy Collier began by identifying the Puget Sound Partnership as a good example of the integration of environmental health and ecosystem services. However, he added that a way to evaluate its effectiveness is needed, particularly to measure any increase in the services produced by Puget Sound, as its protection and restoration have proved very costly.

Collier then referenced a figure from the World Health Organization (WHO) (see Figure 2-4) which illustrates the intersection of climate change and human health, pointing out that though the marine systems and coastal systems were not explicitly called out in the figure, they would be encompassed in the biodiversity loss and ecosystem function and the decline in ecosystem services in general. Collier explained that although the work done to protect coastal ecosystems is focused on providing services to the humans that use them, all aspects of the coastal ecosystem should be considered.

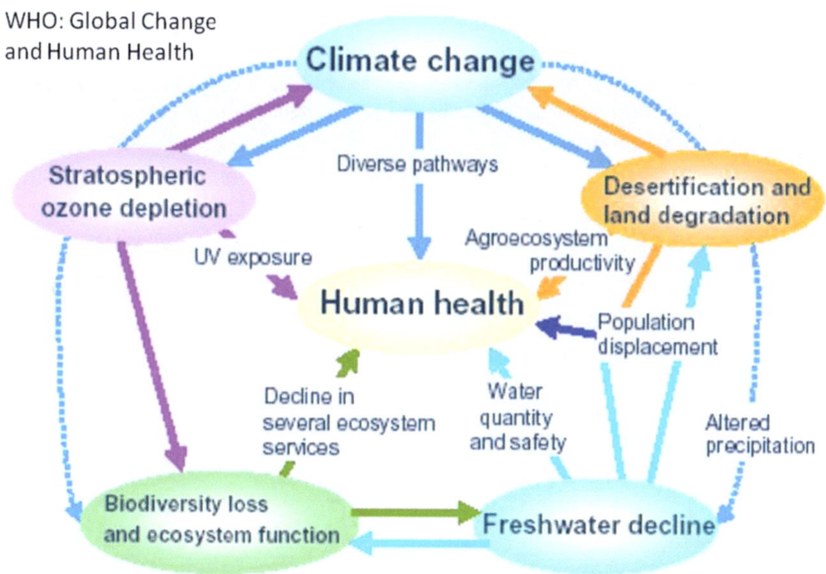

FIGURE 2-4 Linkages between climate change and human health.
SOURCE: WHO, 2013. Reprinted with permission of the World Health Organization.

Much work is being done to identify the stressors that affect the coastal ecosystems, such as nutrients resulting from residential, agricultural, and urban runoff, and consequent harmful algal blooms, particularly cyanobacteria blooms. These stressors have greatly impacted delivery of ecosystem services such as drinking water and recreational access. Though there is a lot of information about stressors and how they affect our ecosystems, Collier noted that there was not very much known about how altered ecosystems actually affect humans.

Puget Sound Partnership

Christine Gregoire, governor of Washington state, created the new state agency, the Puget Sound Partnership, in 2007, to work toward a "swimmable, fishable, diggable Puget Sound by 2020." Collier noted that this defined the political view of ecosystem services for the Partnership; however, in order to put together the policies and write the legislation to create a new state agency, more specific detail was needed. It also became clear that when identifying the types of ecosystem services to be achieved or protected, some kind of translation was necessary to help policy makers understand these terms. After much discussion, six primary goals were identified in order to restore and protect ecosystem services provided by Puget Sound:

- species and food webs (e.g., biological biomass and biodiversity),
- habitats and processes (e.g., eel grass as a habitat that contributes to shoreline protection),[5]
- water quality issues (e.g., toxic contaminants, pathogens, and nutrients in the water),
- water quantity (the annual flows are highly variable, with serious consequences),
- human health, and
- human well-being.

Collier said it was significant that this list explicitly calls out human health and human well-being as specific and separate goals for the Partnership's efforts to restore Puget Sound. However, it is important to find a way to measure improvements in the services and to establish their economic, social, and cultural values.

---

[5] Eel grass softens the impact of waves and currents, preventing coastal erosion and providing a calm space where organic matter and sediments are deposited.

Collier emphasized that salmon is an *extremely* important part of the Puget Sound ecosystem. There are five species of Pacific salmon—chinook, chum, coho, pink, and sockeye. Along with steelhead and bull trout, many populations of all of these anadromous fish species (except pink salmon) are listed under the Endangered Species Act. In addition, there are very substantial legal requirements to maintain salmon populations as part of the tribal treaty rights, which Collier mentioned had been in the news recently, when regional tribes grew dissatisfied with what they perceived as a lack of progress in this area. Unfortunately, there is a great deal of habitat loss due to increasing urbanization, which impacts the salmon life cycle, threatening the juvenile salmon on their path downstream and in nearshore habitats.

Puget Sound is unique among North American estuaries in that it was carved by glacial action and is a deep fjord with shallow sills. This reduces water circulation at several points so that it tends to be isolated hydrologically. Because it is so deep and marine, many species, including some salmon, spend their entire lives within the estuary, which is not the case in most estuaries, where animals come and go at different stages of their lives. It is isolated not only hydrologically, but also, to an extent, biologically. What goes into the Sound, such as toxics, tends not to flush out as happens in other estuaries. A study comparing polychlorinated biphenyls (PCBs) in whole-body chinook salmon along the West Coast found much higher levels in Puget Sound specimens than those to the north or south. These high levels of contaminants have led to fish consumption advisories, limiting Puget Sound salmon consumption. In fact, said Collier, this is a good example of an intersection between the ecosystem services and environmental health, affecting community issues important to policy makers.

Deepwater Horizon Oil Spill and Seafood Safety

Collier next spoke about the Deepwater Horizon explosion of spring and summer 2010. After acknowledging the tragic loss of human life, the fire, and the enormous release of oil into U.S. waters, he turned to the issue of seafood safety. About one-third of the federal fishing waters in the Gulf of Mexico—a major portion of the U.S. fisheries supply, second only to Alaska and the Bering Sea—had to be shut down for several months as a result of the disaster. It takes a long time after an oil spill has closed an area to determine when it can be reopened. Extensive chemical and sensory testing is required, and many hundreds of sites or fish must be sampled to establish the safety of the seafood supply. The testing after the Deepwater Horizon oil spill was both lengthy and expensive.

Aromatic hydrocarbons were presumed to be the most toxic components of oil spills, and fish sampled after the waters had been re-opened were expected to show low or no oil exposure, which federal sampling and analysis confirmed. Indeed, in some cases, levels of contamination were 100 to 1,000 times lower than the government-set levels at which human health is threatened (FDA, 2013). However, Collier noted that, despite these test results, it was very difficult to convince people of the safety of fish taken from the Gulf of Mexico.

Another important aspect of the contamination levels set for safe human consumption of the fish, added Collier, is that fish start to die from the oil exposure at levels far below those set as safe for human consumption. This means that though the scientists knew that there was no chance the fish tested could have been contaminated to the point that eating them would have been a danger to human health—because fish *cannot live* at that the level of contamination, dying at levels far below that—still, the analyses had to be done, to show people the fish were safe. Also, there is a point at which fish will start to smell of oil and, even though these fish would still be considered safe for consumption, they obviously should not be allowed into the market. Collier pointed out that this discrepancy creates a big disconnect between the work of scientists and actual risk to human safety.

The point to consider here, Collier noted, is that most agree that marine fish consumption confers certain health benefits to humans—for example, better cardiovascular health, better development of fetal brains, and perhaps even longer life. So, when a fishery is shut down for an oil spill or for levels of PCBs (as in the Puget Sound fish example above), a large part of what would be an otherwise beneficial food source is being removed from people's diets. In other words, while shutting down fisheries or putting up fish consumption advisories is done to protect human health, there may well be adverse implications for human health as a consequence if the food source cannot be replaced, or is replaced with less healthy food. Collier mentioned that a study is being done on fish consumption in the Gulf of Mexico following the Deepwater Horizon explosion. He expressed a hope that sufficient good data would be collected to show if there was in fact a reduction in rates of fish consumption below normal consumption rates, and if there were a reduction, what connection could be made to any observable health outcomes.

Public Perceptions About Seafood Safety

Collier also discussed the difficulty in balancing public perceptions and communicating about seafood safety. The issue of public fear of fish contamination versus the health benefits of eating fish is a problem in Puget Sound, said Collier. Many people are very concerned about eating salmon (even though the PCB levels may not be health threatening), while on the other hand, there is a population of non-English speakers who need to be warned not to eat bottom fish from severely contaminated areas.

Collier stressed that while the Puget Sound Partnership is cleaning up stormwater and reducing toxic and nutrient inputs—and thus improving the health of the biota, improving the safety of swimming beaches, and reducing harmful algal blooms—they should also be demonstrably improving human health by making access to safe seafood more reliable. Seafood should be regarded by the community as safe and beneficial for both recreational and commercial purposes.

Climate Change Impact on Puget Sound Ecosystem Services

Collier then moved to a consideration of climate change and its impact on ecosystem services as it relates to the loss and modification of salmon habitat. Millions of dollars are currently spent each year in an attempt to get salmon to return in greater numbers to the Sound. Climate change is important in that it affects river flows: projections for 2050 river flows show 30 to 50 percent less water in the streams in Puget Sound during the summer months, which is when many juvenile salmon are rearing in freshwater or making their downstream journey. These reduced water flows during the summer result from reduced snowpacks, in essence changing stream hydrology from snow-driven systems to rain-driven systems. Collier stressed that, in order to maintain ecosystem services in these regional recovery programs, planning for climate change is a necessity.

Indicators to Assess the Health of Coastal and Ocean Environments

The Global Ocean Health Index[6] is a comprehensive measure that scores ocean health. It provides an assessment of the health and benefits of the global ocean, looking at the types of ecosystem services that are

---

[6] The Global Ocean Health Indcx is available at: http://www.oceanhealthindex.org (accessed September 9, 2013).

provided by the coastal oceans (such as food provision, recreation, and so forth). Collier noted that although assigning scores to the various areas is problematic, it is nevertheless a useful exercise, and the data gathered are important. In cooperation with some components of the Puget Sound Partnership, the scientists who created the index will be using their system to assess certain aspects of ecosystem services provided by Puget Sound. Collier pointed out that this system is driven not so much by the state of the ecosystem—as, for instance, biodiversity—but rather is based on what services are provided to humans and is very focused on jobs and other human benefits.

In terms of an ocean health index, Collier noted that the Aral Sea would receive an extremely low score (though, as a freshwater lake, it was not actually included in the index). It was once the fourth largest freshwater lake in the world, but with the diversion of water for farming it suffered a devastating loss of ecosystem services and subsequent great harm to human health and prosperity. Predictions of the oncoming environmental disaster were ignored by the Soviets—the Aral Sea is an example of what can go terribly wrong in ocean management. Nobody wishes anything similar to happen in the United States.

In order to avert such disaster, a first step is to select good environmental indicators to measure the health of the environment. The Puget Sound Partnership has developed a dashboard of 21 vital signs which is used to report the health of Puget Sound and the effectiveness of its actions (see Figure 2-5).

Collier noted that the indicators for the section called "Healthy Human Population" included on-site sewage systems and whether they were properly maintained, and whether swimming beaches and shellfish beds are open or closed (such closures are usually due to fecal contamination). He felt that these indicators were not the best ways of assessing human health. Also, the Puget Sound Partnership has not been successful in building robust indices that could measure quality of life or sound behavior (two of the indicators in the "Human Quality of Life" section) and has not been able to find good data for recreational fishing permits and commercial fishing harvests (the two other indicators in this section). Collier admitted that the Partnership could improve the human health and human well-being aspects of this dashboard system. However, he felt that they were progressing in developing indicators that link ecosystem services to a healthy economy. Work still needs to be done in valuing the cultural and the social aspects, but they are making good progress on economic indicators.

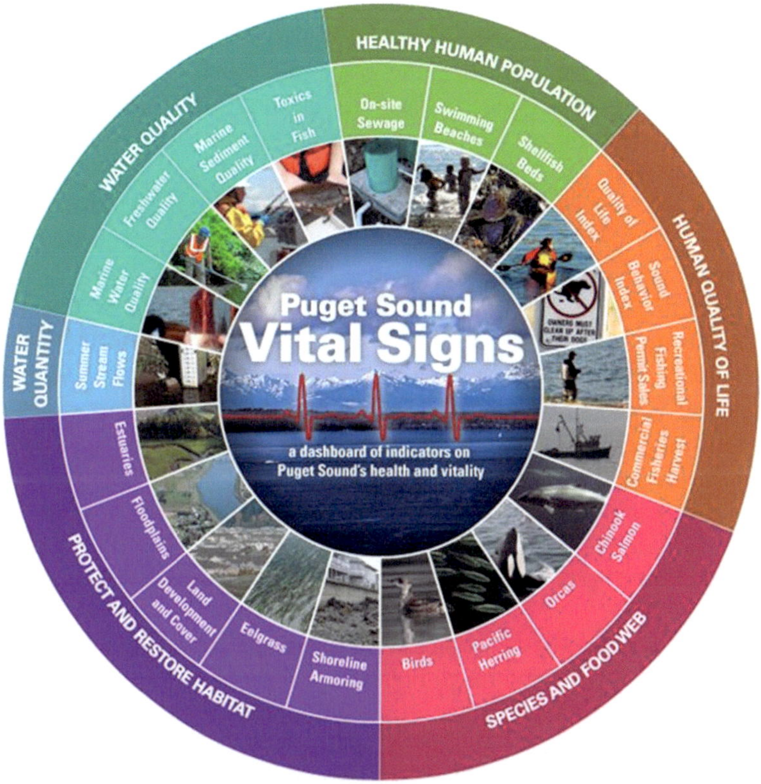

FIGURE 2-5 Puget Sound vital signs wheel.
SOURCE: Puget Sound Partnership, 2012. Vital signs wheel center photo credit: Flickr/Diana K.

Collier concluded his presentation by stressing that better indicators to link coastal and marine condition to ecosystem services and human health and well-being are needed, and asked for any suggestions to further this goal.

DISCUSSION

Lynn Goldman began the discussion by posing the question, what do we need to know to quantify the impacts of ecosystem services on health, especially economically? Olander responded that on the health side there are very good examples of economic valuation. She offered air quality protection through the Clean Air Act as one example where the link between ecosystem services protection and health has been assessed and the economic valuation quantified. She noted that it is more difficult to develop direct health outcomes and economic valuations for ecosystem services that are less concrete such as cultural, spiritual, and mental health but that efforts are being made in developing economic valuations for those services.

Frank Loy commented that quantification is vital because it helps persuade individuals of the importance of the efforts made to protect ecosystem services. He cautioned, however, that quantification of services such as cultural, peaceful, and spiritual must be credible or one could undermine the ability to get public support for those efforts. Olander responded that there are risks with quantifying many things because of uncertainty in a number of parameters. She noted that there is a move away from valuation to evaluation—where stakeholder values are considered rather than economic valuation.

Jack Spengler also addressed the issue of values and how solitude, reflection, and other aspects of how nature is experienced might be quantified. He emphasized that we should not shy away from developing outcomes that are measurable. He mentioned there is work ongoing in Japan and Finland to develop measures for the value of forests, including values that are more difficult to measure. Researchers in these countries have been looking at the restorative functions or therapeutic services provided by forests and at measures such as a reduction in cortisol levels or an increase in serotonin levels. He suggested that the field is a few years away from having the necessary outcomes, but progress is being made in this area.

Christopher Portier commented that the ecosystem services impact on human health is much more difficult to measure, quantify, and value because health outcomes are often the result of a number of factors. For example, chronic diseases are often the culmination of a number of small impacts; it is not a one-exposure and one-disease equation. Olander agreed that this is a problem; current economic valuation is conducted only on those services for which a value can be quantified so most valuations are incomplete and they tend to focus on those with the largest quantifiable benefit.

REFERENCES

Bagstad, K. J., F. Villa, G. W. Johnson, and B. Voigt. 2011. *ARIES—Artificial Intelligence for Ecosystem Services: A guide to models and data, version 1.0.* The Aries Consortium. Available at: http://unstats.un.org/unsd/env accounting/seeaRev/meeting2013/EG13-BG-7.pdf (accessed September 13, 2013).

de Groot, R. S., R. Alkemade, L. Braat, L. Hein, and L. Willemen. 2010. Challenges in integrating the concept of ecosystem services and values in landscape planning, management, and decision making. *Ecological Complexity* 7(3):260–272.

FDA (U.S. Food and Drug Administration). 2013. *Deepwater Horizon oil spill: Questions and answers*. Available at: http://www.fda.gov/Food/Recalls OutbreaksEmergencies/Emergencies/ucm221563.htm (accessed September 9, 2013).

Hanson, C., J. Ranganathan, C. Iceland, and J. Finisdore. 2012. *The corporate ecosystem services review: Guidelines for identifying business risks and opportunities arising from ecosystem change*. Version 2.0. Washington, DC: World Resources Institute.

MEA (Millennium Ecosystem Assessment). 2005. *Ecosystems and human well-being: Synthesis*. Washington, DC: Island Press.

Puget Sound Partnership. 2012. Puget Sound vital signs. Available at: http://www.psp.wa.gov/vitalsigns (accessed September 9, 2013).

Sathirathai, S., and E. B. Barbier. 2001. Valuing mangrove conservation in southern Thailand. *Contemporary Economic Policy* 19(2):109–122.

Sherrouse, B. C., J. M. Clement, and D. J. Semmens. 2011. A GIS application for assessing, mapping, and quantifying the social values of ecosystem services. *Applied Geography* 31(2):748–760.

Tallis, H. T., T. Ricketts, E. Nelson, D. Ennaanay, S. Wolny, N. Olwero, K. Vigerstol, D. Pennington, G. Mendoza, J. Aukema, J. Foster, J. Forrest, D. Cameron, E. Lonsdorf, and C. Kennedy. 2010. *InVEST 1.004 beta user's guide*. Stanford, CA: The Natural Capital Project.

WHO (World Health Organization). 2013. *Global environmental change*. Available at: http://www.who.int/globalchange/climate/en (accessed September 9, 2013).

3

Stressors Impacting Coastal and Ocean Ecosystem Services and Human Health

This chapter provides a summary of presentations outlining stressors that can impact coastal and ocean ecosystem services and possible management decisions and prevention strategies. The first presentation describes how ecosystem stressors—focusing on rising temperatures, eutrophication, ocean acidification, habitat destruction and loss of biodiversity, and extreme weather events—can modify ecosystem services and impact human health. The second presentation describes a framework to allow decision makers to optimize interventions for managing stressors to marine ecosystems in order to maximize the services that will positively impact human well-being. The third presentation provides an agency perspective (U.S. Geological Survey [USGS]) on the role of science in resource management decisions related to ecosystem services. The presentations are followed by a summary of the discussion that ensued.

RELATIONSHIPS AMONG STRESSORS, ECOSYSTEM SERVICES, AND HUMAN HEALTH

Paul A. Sandifer, Ph.D.
Chief Science Advisor, National Ocean Service,
National Oceanic and Atmospheric Administration

Paul A. Sandifer prefaced his remarks by reiterating the four types of ecosystem services described in the Millennium Ecosystem Assessment (MEA, 2005): supporting, provisioning, regulating, and cultural services. All of these services impact human well-being. Health is one specific component of well-being; other components are security, material, and social relations. Together, these contribute to human health and well-being.

Sandifer explained that his presentation is on ecosystem stressors and how those stressors impact changes in ecosystem services, and the ultimate impacts on human health. Health effects, Sandifer said, can be the result of a single stressor, but typically stressors tend to have interacting effects, some antagonistic and some synergistic. However, the greatest likelihood is for stressors to have negative effects on services and ultimately on health outcomes. He pointed out that the presentation would consider the effects of five interacting stressors: rising temperatures, nutrient enrichment, ocean acidification, habitat destruction and its accompanying loss of biodiversity, and extreme weather events and their potential impacts on human health.

Rising Temperatures

Sandifer began the discussion of rising temperatures by suggesting that as the earth's climate warms, heat stress would become a more serious human health problem (see Figure 3-1). Already, health impacts, especially among the elderly, have been seen with more frequent and intense heat waves. In addition to the direct impacts of heat on humans, there are additional impacts of rising temperatures on ecosystems and on the ecosystem services associated with them.

Sandifer described research by Cheung and colleagues (2013), who conducted theoretical studies on more than 600 species of marine fish to evaluate the likely effects of increasing temperatures associated with climate change. Based on the results of their work, the authors suggest that with rising water temperatures, fish populations change in distribution, phenology, and productivity. Average fish body size is also likely to be reduced. According to the study, fish are likely to decrease in size on average by 14 to 24 percent by 2050. This is because global warming will reduce the amount of oxygen in the oceans, and this may also result in dwindling fish catches. Together with overfishing, pollution, and other stresses, these effects may spell additional trouble for the global protein supply in a time of growing need, Sandifer said.

Another impact of rising temperatures relates to food safety. Seafood poisonings are estimated to be underreported, often misdiagnosed, and may be increasing. Pathogenic *Vibrio* is one cause of seafood poisoning. A number of outbreaks in oyster beds in the Gulf of Mexico, New York,

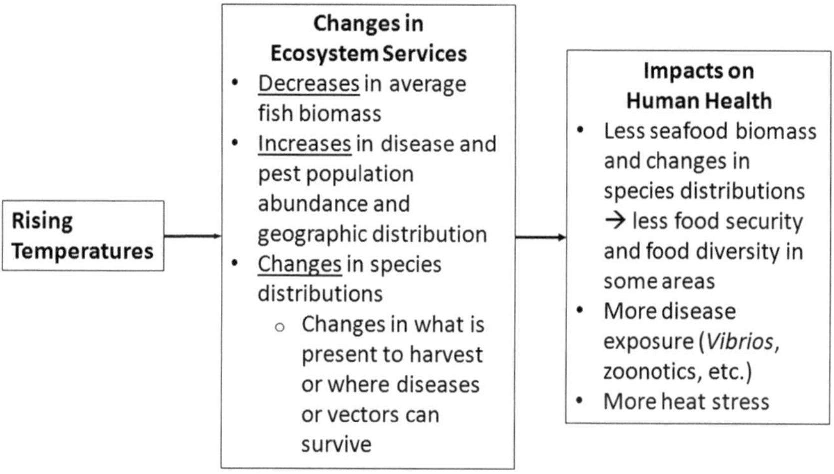

FIGURE 3-1 Anticipated impacts of rising temperatures on ecosystem services and human health.
SOURCE: Sandifer, 2012.

Oregon, and Washington have been associated with *Vibrio parahaemolyticus*. During a period of unusually warm waters in 2004, oyster farms in Prince William Sound, Alaska were devastated by an outbreak of highly virulent *V. parahaemolyticus*. The outbreak resulted in 62 confirmed human cases (McLaughlin et al., 2005) plus others in marine mammals. In the Gulf of Mexico, *V. vulnificus* is a main cause of wound infections among seafood workers. An estimated 200 deaths were attributed to *V. vulnificus* from 1989 to 2004. Illnesses associated with these *Vibrio* species were not required to be reported to the Centers for Disease Control and Prevention (CDC) on a national basis until 2007; thus, it is estimated that the number of infections is likely higher than the CDC annually reports (CDC, 2012). Shellfish beds are typically closed when there is evidence of contamination by *Vibrio* or other infectious organisms. Such closures can seriously undermine public trust in the safety and healthful qualities of seafood.

The distribution and occurrence of zoonotic diseases may also be influenced by temperature and other climate-related factors. Sandifer described two recent examples related to the fungal diseases lacaziosis (previously called lobomycosis) and *Cryptococcus gattii*.[1] Both of these

[1] Lacaziosis a tropical fungal disease typically reported in dolphins and humans. *Cryptococcus gattii* is an uncommon fungal pathogen that affects the lungs and can result in death.

diseases have been found in marine mammals, as well as humans, and appear to be moving northward. Whether this distributional shift is related to climate change is at present unknown, but the diseases are serious and now appear in places and species where they had not been previously seen.

Another zoonotic case of concern Sandifer described was reported by Anthony and colleagues (2012). This case involved harbor seals and avian flu. Between September and December 2011, 162 New England harbor seals died from pneumonia. Postmortem analysis showed the presence of avian influenza virus (H3N8) that was similar to a strain known to be circulating in North American waterfowl since about 2002. The case resulted in a federally recognized unusual mortality event (UME).[2] The authors noted that the outbreak was significant, not only because of the disease it caused in seals, but also because the virus had naturally acquired mutations known to increase transmissibility and virulence in mammals. They emphasized that monitoring the spillover and adaptation of avian viruses in mammalian species is critically important for understanding the factors that lead to both epizootic and zoonotic emergence.

Nutrient Pollution

Sandifer discussed nutrient pollution, another important stressor that causes many problems for coastal and marine environments, including hypoxia and harmful algal blooms (HABs), also known as red tides. Red tides are perhaps the most commonly known HAB, but there are others. Eutrophication, or the overenrichment of water by nutrients such as nitrogen and phosphorus, leads to hypoxia and may increase the occurrence of HABs. These environmental conditions are growing problems worldwide and pose significant human and environmental health risks and can have significant economic impacts, Sandifer said.

Although HABs are most common in coastal and ocean waters, some HABs can occur in the Great Lakes and a variety of other freshwater lakes and ponds. HABs produce potent neurotoxins that can cause a variety of serious illnesses and even death among marine organisms.

---

[2] A UME is defined under the Marine Mammal Protection Act of 1972, Section 404, as a stranding that is unexpected, involves a significant die-off of any marine mammal population, and demands immediate response. See http://www.nmfs.noaa.gov/pr/pdfs/laws/mmpa.pdf (accessed September 9, 2013).

HABs are also toxic to humans, causing a number of illnesses (gastrointestinal, respiratory, neurological, cognitive) and even death. HABs are associated with amnesic shellfish poisoning, ciguatera fish poisoning, diarrheic shellfish poisoning, neurotoxic shellfish poisoning, and paralytic shellfish poisoning. Exposure to HAB toxins can occur via water, seafood (especially filter-feeding molluscan shellfish and some fish), and through aerosols in sea spray. Aerosolized toxins of the Florida red tide *Karenia brevis* have been known to be carried onto beaches and several miles inland where they can cause respiratory system irritation and distress for people, especially those with asthma.

The economic cost of HABs over the past decade has been conservatively estimated at about $1 billion (Jewett et al., 2008). However, the unknown costs in illness, lost productivity, recreational, and other impacts are probably much, much greater, Sandifer said.

HABs pose risks not only for marine life and humans but also for birds and nonmarine mammals. For example, sea birds may be affected by eating HAB-contaminated shellfish or fish (Landsberg et al., 2009). At one point during 2009, officials recommended that visitors not bring dogs to the Padre Island National Seashore as dogs and coyotes had become ill or died possibly as a result of consuming fish that had been killed by a red tide, Sandifer said.

Sandifer noted that there is growing evidence for multiple species of HABs and geographic areas, suggesting that a warming climate will result in increased frequency, duration, and geographic extent of HABs (Gilbert et al., 2005; Van Dolah, 2000). For example, Moore and colleagues (2008, 2011) estimated that the window of opportunity for blooms of the toxic alga *Alexandrium catanella*, which produces saxitoxin, a paralytic toxin, will increase by 13 days on average, begin up to 2 months earlier, and will persist for up to an additional month by the end of the century. In other work, Sun and colleagues (2011) found that increases in dissolved carbon dioxide and reduced phosphate levels, as observed in ocean waters, can increase growth of the toxic diatom *Pseudo-nitzschia* and production of its toxin, domoic acid, which causes amnesic shellfish poisoning in humans.

Sandifer noted that recent coastal surveys of the United States and Europe found that 78 percent of the assessed continental U.S. coastal area and approximately 65 percent of Europe's Atlantic coast exhibit symptoms of eutrophication and the problem is growing at alarming rates (Diaz and Rosenberg, 2008). In the United States the three largest hypoxic zones are the Gulf of Mexico, the Chesapeake Bay, and Lake

Erie. In general, nutrient pollution of coastal areas is an important issue and may impact coastal ecosystems services in many ways, including decreases in clean water, safe food, breathable air, and coastal recreational opportunities, Sandifer said.

Ocean Acidification

Sandifer observed that ocean waters are now 30 percent more acidic than preindustrial levels (NOAA, 2013). This increased acidity is having negative effects on ocean organisms. Sandifer described a recent meta-analysis (Kroeker et al., 2010) that found ocean acidification to have negative effects on survival, calcification, growth, and reproduction of a variety of marine organisms. The study also found significant variation in the sensitivity of marine organisms. Calcifying organisms (e.g., corals, mussels, phytoplankton) generally exhibited larger negative responses than noncalcifying organisms across numerous response variables, with the exception of crustaceans, which calcify but were not negatively affected. Corals, in particular, are negatively affected by ocean acidification. Molluscan shellfish also exhibit a negative response to ocean acidification, which threatens the availability and economic benefit of this type of seafood. The aesthetic benefits of coral reefs and the ecotourism opportunities are also affected by ocean acidification.

Sandifer provided an example of the impact of ocean acidification on a specific ecosystem service, the production of farmed oysters in the Pacific Northwest. Oyster hatcheries in this area that were on the verge of collapse a few years ago are now again major contributors to the West Coast shellfish industry, he said. Beginning in 2005, production at some Pacific Northwest oyster hatcheries began declining at an alarming rate, posing a severe economic impact and challenging a way of life held by shellfish growers for more than 130 years (Washington State Blue Ribbon Panel on Ocean Acidification, 2012). Oyster production represents about $84 million of the West Coast shellfish industry and supports more than 3,000 jobs. A $500,000 congressional investment in monitoring the pH of coastal seawater, which enables hatchery managers to schedule production when water quality is good, is helping to restore commercial hatcheries.[3] However, much more work, including continued monitoring, is needed to help safeguard the ongoing contribution of this important

[3] See http://www.noaa.gov/features/01_economic/pacificoysters.html (accessed September 9, 2013).

industry to coastal communities in Oregon and Washington. This example highlights the urgency of this problem and the value of ocean acidification research and monitoring (Barton et al., 2012).

Sandifer highlighted that, in addition to negative effects on shellfish production, the health of coral reefs, and potentially food and economic security, the stress of dealing with ocean acidification means that coastal ecosystems may become less resilient to other stressors, including extreme weather, nutrient pollution, or overfishing, becoming less able to recover from these types of challenges.

Habitat Destruction and Biodiversity Loss

The fourth example Sandifer discussed was habitat destruction and biodiversity loss. Coastal habitats are some of the most threatened in the world. Most of this loss is due to sea-level rise or coastal development, Sandifer said. As these systems are lost, biodiversity and many ecosystem services are lost. For example, oyster reefs provide many services, including seafood, filtration services, and water quality benefits, as well as shoreline protection and stabilization. Similarly, coral reefs provide many ecosystem services, including food, medicines and other products, nursery habitat for other species, and recreational opportunities. Healthy dunes and beaches confer storm protection for shorelines and human habitations, among other services.

Degradation and loss of natural coastal habitats and biodiversity results in diminished storm surge protection, seafood supply, nutrient processing, and recreational and aesthetic values and generally decreases the resilience of coastal ecosystems to other stressors. The cumulative effect is greater risk of property damage and loss of life during storms, less seafood, fewer jobs and reduced food security, more risks of water-related illness, and impacts to mental health, Sandifer said.

Extreme Weather Events

Sandifer noted that degraded coastal habitats often exacerbate impacts of extreme weather, and extreme events often degrade or destroy coastal habitats and the ecosystem services they produce. Such effects were seen with Hurricane Katrina, in the aftermath of the Deepwater Horizon oil spill, and with posttropical storm Sandy, Sandifer said. He reminded the audience that in 2011 the United States experienced 14 weather and climate disasters, each of which exceeded $1 billion in losses (Smith and Katz, 2013). These ranged from winter blizzards that affected two-thirds of the United States (from Texas and Oklahoma to New England) early in the year, to a record number of tornadoes in the spring in the Midwest and Southeast (including the deadliest one to date, which killed 160 people), spring flooding in the Mississippi, record wildfires in the West, and Hurricane Irene late in the summer, which produced flooding. The flooding in the Mississippi produced near-record hypoxia in the Gulf of Mexico as well (NOAA, 2011).

In addition to the lives that are lost due to storms and other extreme weather events, these events also often have a number of impacts on ecosystems and on their ability to provide ecosystem services. For example, storms may damage coastal habitats, causing a loss of coastal habitats and loss of future storm surge protection. Storms also frequently overwhelm sewer systems (particularly combined sewer systems with sewage and stormwater in the same system), resulting in contamination of drinking water and affecting recreational water use (Patz et al., 2010; Portier et al., 2010).

Storms also damage human infrastructure, leading to leaks of pollutants which contaminate ecosystems. For example, there was concern over the potential for seafood contaminated by radiation from the Fukushima nuclear power plant that was damaged by the major March 2011 earthquake and tsunami (Reardon, 2011). Contamination of coastal ecosystems with pollutants can decrease water quality, affect seafood safety (radiation, oil, sewage), and contribute to losses of wildlife, and recreational opportunities.

Sandifer also noted that, in addition to the direct impacts of weather events on human health, the aftermath of such occurrences may include delayed effects on human and environmental health. For example, infections, illnesses, and mental health issues may arise following weather events, and dealing with these may be complicated by an inability to get medications or medical care due to storm damage to medical or transportation infrastructure.

Sandifer noted that the five environmental stressors he discussed are interrelated and all have effects on the provision of ecosystem services, which in turn can have a broad range of effects on human health (see Figure 3-2). He suggested several steps that could be taken to address the stressors and sustain marine and coastal ecosystem services. These include the following:

- explicitly account for ecosystem services in policies and decision making;
- protect and restore coastal "green infrastructure" (i.e., intact coastal habitat) to provide natural storm surge protection, food security, and climate adaptation benefits;
- conduct research to understand the effects of environmental stressors on species, habitats and systems, and humans so we can determine how best to mitigate and adapt; and
- implement better monitoring and health warning systems.

In concluding his presentation, Sandifer said that the complex interactions among multiple stressors, ecosystem services, and human health highlight the need to more fully understand the connections among these factors and their ultimate human health impact so that the impacts can be minimized. He provided an example of the time scales at which climate and weather effects are considered. The National Oceanic and Atmospheric Administration (NOAA) develops long-term outlooks for climate and weather and then refines these to finer and finer time scales, down to days, hours, and minutes of actual forecasts of impending events, so that preventive or protective actions can be taken with the greatest lead time possible (see Figure 3-3).

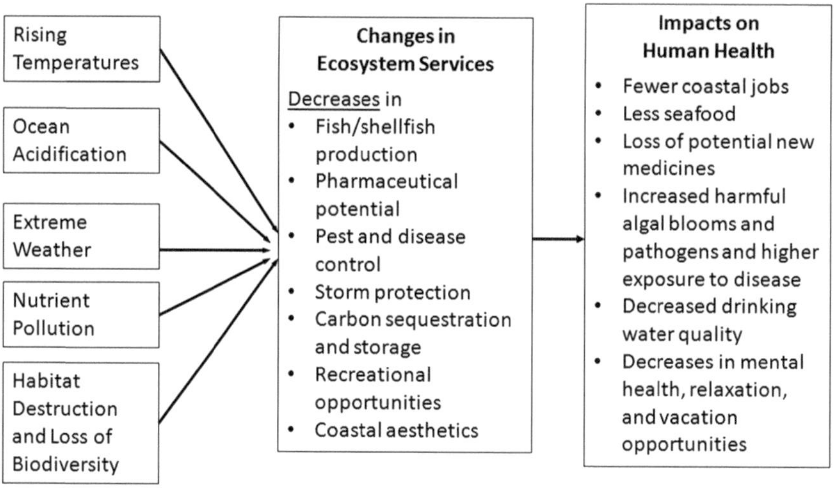

FIGURE 3-2 Interactions among five environmental stressors, delivery of ecosystem services, and impacts on human health.
SOURCE: Sandifer, 2012.

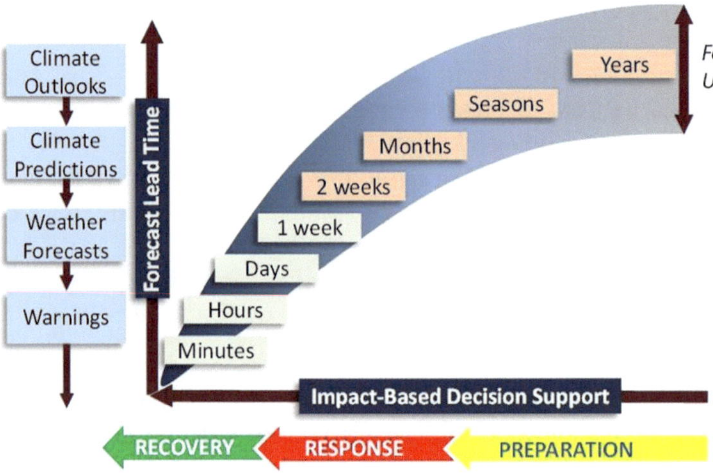

FIGURE 3-3 Time scales and types of climate and weather forecasts.
SOURCE: Sandifer, 2012.

NOAA is now extending this type of approach in an agency-wide effort to improve ecological forecasts. Initially, the agency is focusing on harmful algal blooms, hypoxia, and pathogens (particularly naturally occurring *Vibrio* bacteria) in coastal and marine environments. Sandifer noted that capability and resources (including those for disease and health surveillance and epidemiological studies—to monitor, integrate data, model, and forecast impacts to coastal and ocean ecosystem services and the resulting human health threats) are needed in order to provide timely warnings that would enable better preparation and mitigation, implementation of control and prevention strategies, reduction of impacts, and shortened recovery times.

FRAMEWORK FOR ASSESSING MARINE ECOSYSTEM SERVICES AND HUMAN HEALTH

Jonathan Garber, Ph.D.
Acting Associate Director for Ecology,
National Health and Environmental Effects Research Laboratory,
U.S. Environmental Protection Agency

Jonathan Garber began his presentation by explaining that enhancing and protecting ecosystems and human health are explicitly central to the mission of the U.S. Environmental Protection Agency (EPA) and to the National Health and Environmental Effects Research Laboratory. The EPA's efforts in the ecosystem–health connection area of research are a continuing and increasing focus of the agency, he said. The work is supported by both the regulatory and science arms of the agency and is underpinned by a number of statutory authorities such as the Clean Water Act, the Marine Protection Research and Sanctuaries Act, the Ocean Dumping Ban Act, the Shore Protection Act, and others.

Decision Framework

Garber described a process used by the EPA to assess the linkage between ecosystems and human health. The process is based upon a framework that was developed to assist in optimizing management decisions that affect the production of coastal and marine and ocean ecosystem services. The framework explicitly includes linkage to human well-being.

As can be seen in Figure 3-4, the framework begins on the left side with potential stressors on marine systems such as temperature rise, eutrophication, and habitat loss. The next step is to consider the potential management decisions and points of intervention in the delivery of ecosystem services and how they affect the production of these services. At the far right of the schematic is the linkage to the desired outcome of human well-being. Also of note is the understanding that there are important feedbacks loops. This framework allows a decision maker to optimize decisions about potential interventions in order to maximize the production of services that will have a positive impact on human well-being.

In support of this framework, Garber discussed a critical path for making the connection between ecosystem services and human health outcomes. The first step in the path is to establish inventories of baseline ecosystem services and health conditions. The second step is to translate conditions into quantifiable services. The third step is to link these services to human health outcomes, and the fourth step is to model and predict the impacts of interventions and feedbacks to complete the cycle. As an example, Garber described the use of this path in assessing coastal conditions and human health. The results of the first step to establish inventories of baseline ecosystem services and health conditions are captured in the National Coastal Assessment series of reports.

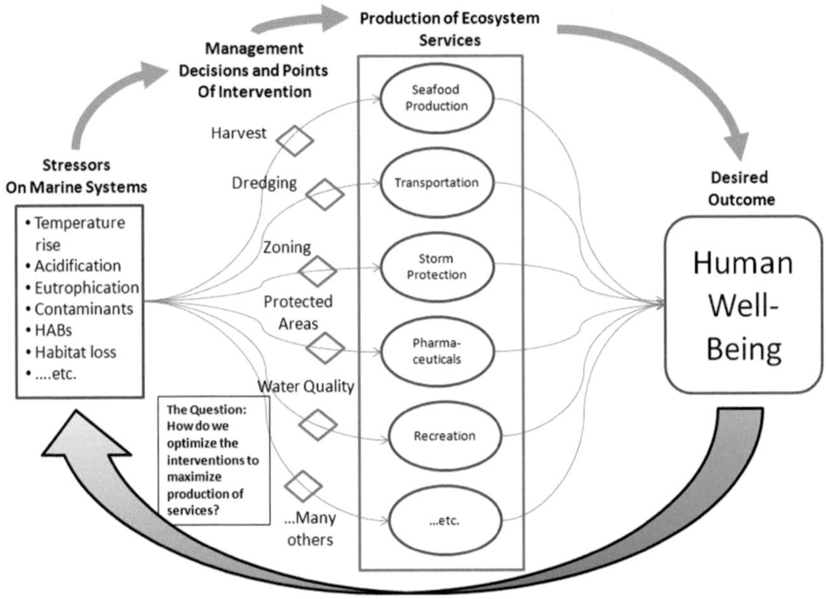

FIGURE 3-4 Modified DPSR (drivers, pressures, states, response) decision framework for optimizing management decisions that affect production of coastal marine and ocean ecosystem services.
NOTE: HABs = harmful algal blooms.
SOURCE: Garber, 2012.

National Coastal Condition Assessment

The National Coastal Assessment began in 2000 as an integrated comprehensive coastal monitoring program to assess the condition of estuaries at multiple scales (state, regional, and national). The program included all U.S. coastal states. Another important aim was to transfer this technology to the states, tribes, EPA Regions, EPA Office of Water, and others, and to enhance the EPA's ability to make scientifically sound assessments of the condition of U.S. coastal waters. This effort ended in 2006 and was replaced by the National Coastal Condition Assessment (NCCA). Four NCCA reports have been published.

According to Garber, the assessments have evolved over the past 20 years and now include all the coterminous U.S. coastal waters, particularly estuaries, and are now reaching out to the continental shelf and some of the territories and states that are not coterminous. Data contained in the last report, *NCCR IV*, includes coastal monitoring data, offshore fisheries data, coastal ocean data, and assessment and advisory data related to fish consumption advisories and beach closures. Data included in the report are indicator based; an illustrative set of indicators is shown in Box 3-1.

Synthesis of this information allows for a national assessment to be made of the condition of U.S. coastal waters. The *NCCR IV* assessment results rated U.S. coastal waters as "fair" based on a five-point system, where a score of less than 2.0 is rated poor, and greater than 4.0 is rated good. The water quality, sediment quality, benthic condition, and coastal habitat indices also rated fair. The fish tissue contaminants index rated "good to fair" (>3.7–4.0). Regional chapters of the *NCCR IV* provide information on indicators for regional coastal areas.

EnviroAtlas and Eco-Health Browser

Garber discussed another EPA tool, the EnviroAtlas,[4] which is useful in the second step of the framework, translating conditions to quantifiable

BOX 3-1
Indicators of Coastal Condition in *NCCR IV* Report

Water Quality Index:
- Water clarity
- Dissolved oxygen (DO)
- Dissolved inorganic nitrogen (DIN)
- Dissolved inorganic phosphorus (DIP)
- Chlorophyll a

Benthic Index[a]:
- Community diversity
- Pollution-tolerant/sensitive species

Sediment Quality Index:
- Toxicity
- Contaminants
- Total organic carbon (TOC)

Fish Tissue Contaminant Index:
- Whole-fish contaminant burden

Coastal Habitat Index:
- Fish and Wildlife Service National Wetlands Inventory Wetlands Loss Rates

[a] Benthic index measures the condition of the benthic organisms living in or on the bottom of water bodies.
SOURCE: EPA, 2011.

---

[4] EnviroAtlas is available at: http://www.epa.gov/research/enviroatlas/index.htm (accessed September 9, 2013).

services. The Web-based tool allows research and analysis to be conducted on the relationships between ecosystem services and human well-being. The tool provides easy-to-use geospatial data and maps that allow users to analyze multiple ecosystem services and health conditions in a specific region. These ecosystem benefits include clean air, clean and plentiful water, natural hazard mitigation, biodiversity conservation, food, fuel, materials, recreational opportunities, and cultural and aesthetic value. EnviroAtlas contains information on the status of these benefits, the ecosystems that provide and protect them, and related health and economic impacts.

Another tool, the eco-health browser,[5] is particularly useful in conducting step three: linking services to human health outcomes. The interactive tool provides information on major ecosystems (e.g., wetlands and forests), the services they provide, and how those services or their degradation and loss may affect human health (see Figure 3-5).

Garber ended his presentation by sharing his view that much progress has been made in establishing the baseline of ecological and human health conditions, in translating conditions to quantifiable services, and in linking services to human health outcomes. However, much more work is needed in step four of the framework: modeling and predicting the impacts of interventions and feedbacks. Progress in this area would allow for completing the cycle of optimizing management decisions that affect the production of coastal marine and ocean ecosystem services.

A U.S. GEOLOGICAL SURVEY PERSPECTIVE ON STRESSORS IMPACTING COASTAL AND OCEAN SYSTEMS

Ione Taylor, Ph.D.
Associate Director, Energy and Minerals,
and Environmental Health,
U.S. Geological Survey

Ione Taylor began her presentation by describing the mission of the USGS and how the agency's work addresses ecosystem services and its impact on health. The USGS is a science agency within the U.S. Department of the Interior. The USGS was founded in the late 1800s to classify

---

[5] The eco-health browser is available at: http://www.epa.gov/research/healthscience/browser/index.html (accessed September 9, 2013).

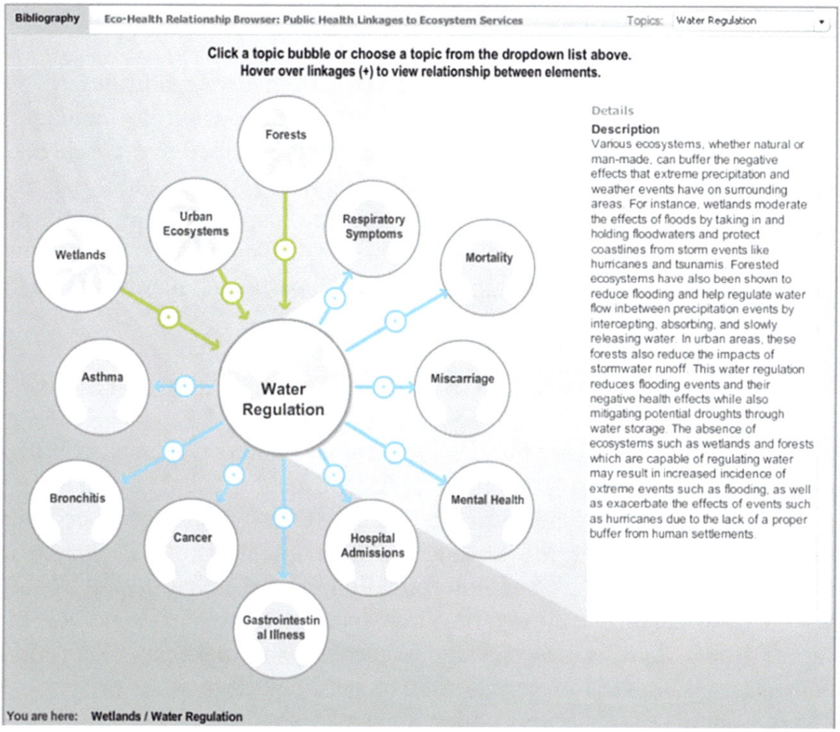

FIGURE 3-5 Eco-Health Relationship Browser.
NOTE: The Eco-Health Relationship Browser is best viewed in interactive format at: http://www.epa.gov/research/healthscience/browser (accessed September 9, 2013).
SOURCE: EPA, 2013.

the public lands and examine the geological structure, mineral resources, and products of the national domain. The mission of the agency is to provide reliable scientific information to describe and understand the Earth; minimize the loss of life and property due to natural disasters; manage water, biological, energy, and mineral resources; and enhance and protect quality of life. Taylor noted that 2 years ago the USGS underwent a reorganization to move from a long history of functioning as a discipline-based organization with an academic structure to an issue-based structure of Mission Areas. Ecosystem services is now a cross-cutting issue across all Mission Areas. One area of USGS focus on ecosystem services is on valuation and how USGS biophysical science can contribute to understanding value, particularly in the discussion of trade-offs for natural resource management and land management.

Ecosystem Responsibilities of the Department of the Interior

As background, Taylor described ecosystem responsibilities of the Department of the Interior. The Department is responsible for managing 35,000 miles of coastline, 80 billion acres of the seabed and continental shelf and subsurface minerals, 177 island coastal refuges, and a fish and wildlife refuge system. The Department is also responsible for 74 marine or island national parks and 92 million acres of coral reef ecosystems. The Department thus has an important role in coastal and marine oversight for the nation.

The Role of Science in USGS Resource Management Decisions

Taylor also noted that the USGS is not a regulatory agency. It brings unbiased science research to the Department's land management responsibilities. She pointed out that the usefulness of science to decision makers for making high-level land management decisions is related to the degree of synthesis and interpretation of the science. Figure 3-6 shows how science informs resource management decisions. As the figure shows, there is greater value to the decision maker as biophysical data are transformed to information that can be used to make predictions and present potential options within an ecosystems services framework.

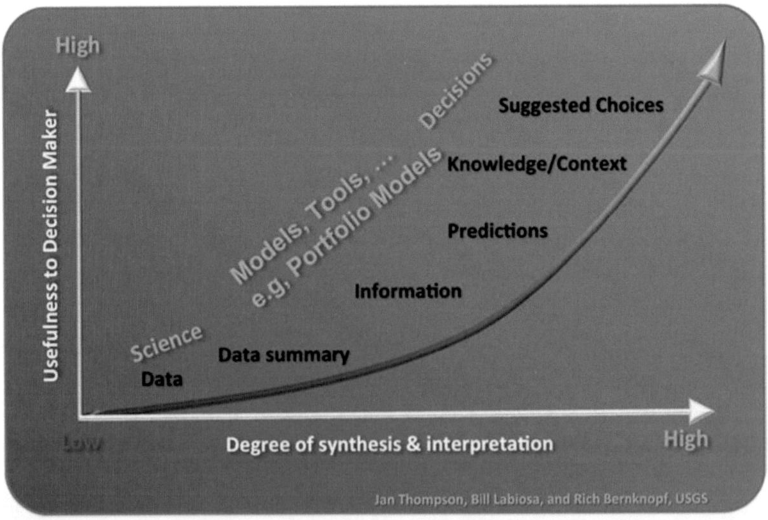

FIGURE 3-6 How science informs resource management decisions.
SOURCE: Taylor, 2012.

To support this effort, the USGS is expanding its basic core capabilities in science to develop a capacity to better integrate information to inform complex trade-offs and decisions, Taylor said. Figure 3-7 shows the USGS's approach to move from the traditional core capabilities to emerging capabilities. The Energy and Mineral Resources Mission Area is a traditional USGS core capability and is shown on the left. The core capability of energy and mineral appraisal is assessed for the nation and globally because the future supply of minerals, and to a great extent energy, is in the global import arena. The USGS has long-standing capability in economic modeling and the ability to conduct appraisals and valuation for these commodities. Over the past 10 to 15 years the USGS has begun to bring environmental impacts, particularly of energy and minerals and water resources, to bear for a broader picture of environ-mental impact for resource appraisal, potential extraction scenarios, and transport and use. Most recently an emerging capability is the development of biophysical models in an ecosystems services framework. Ultimately the goal is to move toward building the capability to generate scenarios for vulnerability and risk optimization in complex natural resource and land management decisions. This requires taking core capabilities in the biophysical sciences, building in what is known about economic appraisals and services in the energy and mineral area, and bringing a similar perspective into a consideration of the value of biologic components and hydrologic components, which typically have not had a commodity or price. Expanding the core capabilities will require more synthesis and more partnership efforts with other agencies such as the EPA and NOAA.

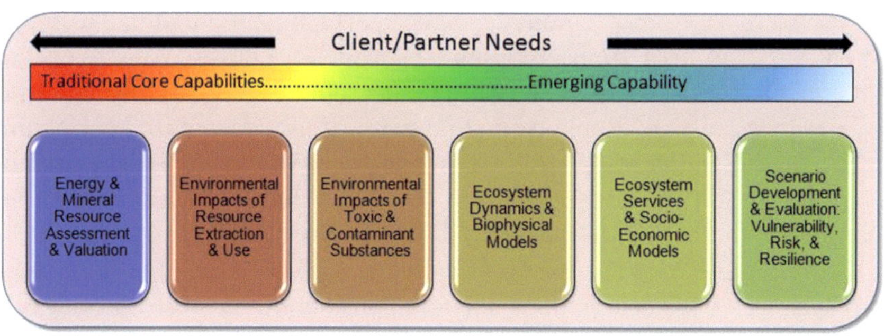

FIGURE 3-7 A developing USGS approach: integrating traditional and emerging capabilities.
SOURCE: Taylor, 2012.

The Role of the USGS in Addressing Sea-Level Rise, Methylmercury, and Natural Hazards

Taylor provided a brief overview of three stressors that impact coastal and ocean ecosystem services: (1) sea-level rise related to climate and coastal erosion; (2) environmental health related to water quality, disease, mercury contamination, and ocean acidification; and (3) natural hazards such as earthquake, tsunami, and landslide hazard stressors.

In the area of sea-level rise the USGS is taking current knowledge based on the geologic and geoscience perspective and the hydrologic perspective to develop high-level national and sometimes international assessments. It is also conducting research on sea-level rise, coastal erosion, fragile shorelines in coastal areas, and wetland vulnerabilities. The USGS is bringing the science to bear to understanding the interplay of sea-level rise, coastal erosion, and a host of other drivers that impact coastal vulnerability and is developing different scenarios of sea-level change to assess impact. Communicating these assessments to the public is a challenge.

Taylor turned to the issue of environmental health and described the USGS's work on methylmercury. The first national map of methylmercury based on a model of surface waters was recently completed by the USGS. In the United States the primary source of mercury over the land mass comes from deposition of ash from coal-fired power plants. Mercury becomes a problem for human environmental health because it becomes toxic through methylation of the mercury, which is a biomediated process. That is, it is the combination of the microbes in the water and mercury that results in toxicity. The USGS is looking at the problem of methylmercury from a regional perspective. One region that has received attention is an area from Honolulu, Hawaii, to Kodiak, Alaska, in the Pacific Ocean. Results of work on this topic in the Pacific indicate concerns for a large plume of methylmercury in the oceans. That plume is associated with atmospheric deposition from coal-fired power plants in Asia which is being deposited into the water due to interactions at the land–sea interface.

Taylor emphasized that an important finding from this and other work is that assessments must include consideration of the system as a whole and this requires an interdisciplinary approach to science. This approach is difficult because it requires translating across science cultures, science practice, language, data, and how data are collected.

Taylor described the unique role of the USGS with respect to natural hazards. The USGS has delegated responsibility for the federal government to provide notifications and warnings for earthquakes, volcanic eruptions, and landslides. The agency also works to support NOAA's flood and severe weather (including hurricane) warnings, as well as seismic networks to support tsunami warnings. The agency's work around natural hazards such as earthquake and tsunami warnings and landslides is also important with respect to work and concerns related to sea-level rise.

Taylor concluded her presentation by presenting six questions and ongoing challenges around the topic of stressors and ecosystem services for coastal and marine ecosystems:

1. How can ecosystem services and their values be routinely incorporated into sustainable marine resource management decisions so that the impacts of decisions across natural, managed, and human systems are understood?
2. What scientific information and what level of certainty are needed to provide a foundation for coastal and marine resource management decisions?
3. How can the value of scientific information be more effectively determined and used to prioritize the science needed to better understand coastal and marine ecosystem services?
4. What metrics should be developed so that we understand the ability of marine systems to recover from expected and unexpected stressors?
5. How can the natural and social sciences collaborate to develop integrated understanding of the consequences expected to result from stressors and management decisions?
6. How can marine ecosystem services be incorporated into adaptive decision making that facilitates the synthesis of learning and management?

DISCUSSION

A brief discussion followed the panelists' presentations. Lynn Goldman stated that there has been much scientific discussion about climate change and the impact of temperature rise, and the role of human activities on greenhouse gases, but yet as a nation we are not managing ecosystems in a holistic fashion, or including humans as a part of that equation, and we are therefore also not preparing for changes related to climate. She asked the panel whether part of the problem with a lack of urgency in better managing ecosystems is that the knowledge available is not being translated, or whether there is missing knowledge, or whether the message of ecosystem change is not well communicated to policy makers and the public.

All of the panelists responded that improving communication to policy makers and the public is critical. Sandifer noted that improved communication also needs to occur among sectors and the public. Sandifer suggested that social media could be harnessed to extend the reach and persistence of messaging on this topic to the public. Garber responded that we need to create an environmental ethic; there are a myriad of steps we can take in our personal lives to protect ecosystems, but we need to create an environmental ethic.

Christopher Portier commented that agencies have their own cultures and they tend to look at issues from their specific perspective. What is needed, Portier said, is a systems perspective. He suggested that perhaps it is time to reassess our policy and regulatory frameworks which were developed decades ago and move toward a holistic, systems perspective.

REFERENCES

Anthony, S. J., J. A. St. Leger, K. Pugliares, H. S. Ip, J. M. Chan, Z. W. Carpenter, I. Navarrete-Macias, M. Sanchez-Leon, J. T. Saliki, J. Pedersen, W. Karesh, P. Daszak, R. Rabadan, T. Rowles, and W. I. Lipkin. 2012. Emergence of fatal avian influenza in New England harbor seals. *mBio* 3(4):e00166-12, doi:10.1128/mBio.00166-12.

Barton, A., B. Hales, G. G. Waldbusser, C. Langdon, and R. A. Feely. 2012. The Pacific oyster, *Crassostrea gigas*, shows negative correlation to naturally elevated carbon dioxide levels: Implications for near-term ocean acidification effects. *Limnology and Oceanography* 57(3):698–710.

CDC (Centers for Disease Control and Prevention). 2012. *Cholera and other Vibrio illness surveillance overview*. Atlanta, GA: U.S. Department of Health and Human Services.

Cheung, W. W. L., J. L. Sarmiento, J. Dunne, T. L. Frölicher, V. W. Y. Lam, M. L. Deng Palomares, R. Watson, and D. Pauly. 2013. Shrinking of fishes exacerbates impacts of global ocean changes on marine ecosystems. *Nature Climate Change* 3:254–258.

Diaz, R. J., and R. Rosenberg. 2008. Spreading dead zones and consequences for marine ecosystems. *Science* 321(5891):926–929.

EPA (U.S. Environmental Protection Agency). 2011. *National Coastal Condition Report IV.* Washington, DC: EPA. Available at: http://www.fws.gov/wetlands/Documents/National-Coastal-Condition-Report-IV-part-1-of-2.pdf (accessed September 9, 2013).

EPA. 2013. EnviroAtlas: Eco-Health Relationship Browser. Available at: http://www.epa.gov/research/healthscience/browser/index.html (accessed September 9, 2013).

Garber, J. 2012. *Framework for assessing marine ecosystem services and human health.* Presentation at the Institute of Medicine Workshop on Understanding the Connections Between Coastal Waters and Ocean Ecosystem Services and Human Health: Basic Services, Valuation, and Resiliency. Washington, DC.

Gilbert, P. M., D. M. Anderson, P. Gentien, E. Graneli, and K. G. Sellner. 2005. The global complex phenomena of harmful algae. *Oceanography* 18(2):136–147.

Jewett, E. B., C. B. Lopez, Q. Dortch, S. M. Etheridge, and L. C. Backer. 2008. *Harmful algal bloom management and response: Assessment and plan.* Interagency Working Group on Harmful Algal Blooms, Hypoxia, and Human Health of the Joint Subcommittee on Ocean Science and Technology. Washington, DC.

Kroeker, K. J., R. L. Kordas, R. N. Crim, and G. G. Singh. 2010. Meta-analysis reveals negative yet variable effects of ocean acidification on marine organisms. *Ecology Letters* 13(11):1419–1434.

Landsberg, J. H., L. J. Flewelling, and J. Naar. 2009. *Karenia brevis* red tides, brevetoxins in the food web, and impacts on natural resources: Decadal advancements. *Harmful Algae* 8(4):598–607.

McLaughlin, J. B., A. DePaola, C. A. Bopp, K. A. Martinek, N. P. Napolilli, C. G. Allison, K. Shelley, E. C. Thompson, M. M. Bird, and J. P. Middaugh. 2005. Outbreak of *Vibrio parahaemolyticus* gastroenteritis associated with Alaskan oysters. *New England Journal of Medicine* 353(14):1463.

MEA (Millennium Ecosystem Assessment). 2005. *Millennium ecosystem assessment. Synthesis report.* Washington, DC: Island Press.

Moore, S. K., V. L. Trainer, N. J. Mantua, M. S. Parker, E. A. Laws, L. C. Backer, and L. E. Fleming. 2008. Impacts of climate variability and future climate change on harmful algal blooms and human health. *Environmental Health* 7(Suppl. 2):S4, doi:10.1186/1476-069X-7-S2-S4.

Moore, S. L., N. J. Mantua, and E. P. Salathé. 2011. Past trends and future scenarios for environmental conditions favoring the accumulation of paralytic shellfish toxin in Puget Sound shellfish. *Harmful Algae* 10(5):521–529.

NOAA (National Oceanic and Atmospheric Administration). 2011. *NOAA-supported scientists find large dead zone in Gulf of Mexico.* Available at: http://www.noaanews.noaa.gov/stories2011/20110804_deadzone.html (accessed September 9, 2013).

NOAA. 2013. *State of the science fact sheet: Ocean acidification*. Washington, DC: NOAA.

Patz, J. A., S. J. Vavrus, C. K. Uejo, and S. L. McLellan. 2010. Climate change and waterborne disease risk in the Great Lakes region of the U.S. *American Journal of Preventive Medicine* 35(5):451–458.

Portier, C. J., K. Thigpen Tart, S. R. Carter, C. H. Dilworth, A. E. Grambsch, J. Gohlke, J. Hess, S. N. Howard, G. Luber, J. T. Lutz, T. Maslak, N. Prudent, M. Radtke, J. P. Rosenthal, T. Rowles, P. A. Sandifer, J. Scheraga, P. J. Schramm, D. Strickman, J. M. Trtanj, and P.-Y. Whung. 2010. A human health perspective on climate change: A report outlining the research needs on the human health effects of climate change. *Environmental Health Perspectives* doi:10.1289/ehp.1002272.

Reardon, S. 2011. Fukushima radiation creates unique test of marine life's hardiness. *Science* 332(6027):292.

Sandifer, P. 2012. *Stressors impacting coastal and ocean ecosystem services and human health*. Presentation at the Institute of Medicine Workshop on Understanding the Connections Between Coastal Waters and Ocean Ecosystem Services and Human Health: Basic Services, Valuation, and Resiliency. Washington, DC.

Smith, A. B., and R. W Katz. 2013. U.S. billion-dollar weather and climate disasters: Data sources, trends, accuracy, and biases. *Natural Hazards* 67(2):387–410.

Sun, J., D. A. Hutchins, Y. Feng, E. L. Seubert, D. A. Caron, and F.-X. Fu. 2011. Effects of changing pCO_2 and phosphate availability on domoic acid production and physiology of the marine harmful bloom diatom *Pseudo-nitzschia* multiseries. *Limnology and Oceanography* 56(3):829–840.

Taylor, I. 2012. *A U.S. Geologic Service (USGS) perspective on stressors impacting coastal and ocean systems*. Presentation at the Institute of Medicine Workshop on Understanding the Connections Between Coastal Waters and Ocean Ecosystem Services and Human Health: Basic Services, Valuation, and Resiliency. Washington, DC.

Van Dolah, F. M. 2000. Marine algal toxins: Origins, health effects, and their increased occurrence. *Environmental Health Perspectives* 108(Suppl. 1):133–141.

Washington State Blue Ribbon Panel on Ocean Acidification. 2012. *Ocean acidification: From knowledge to action, Washington State's strategic response*. Publication no. 12-01-015, edited by H. Adelsman and L. W. Binder. Olympia, WA: Washington Department of Ecology.

4

Seafood Supplies and Food Security

Seafood is an important ecosystem service provided by coastal waters and ocean ecosystems and is a primary source of protein for close to a billion people. The first presentation in this chapter addresses the need to balance human dietary needs and the sustainability of fisheries and the ecosystems that provide them. The second presentation further elaborates on the need to identify and implement responsible strategies for managing ecosystems that provide fish and other seafood. Aquaculture is discussed in the third presentation as a strategy to increase seafood availability and safety. The fourth presentation discusses the use of life-cycle analysis in assessing decision making related to sustainable fisheries. Each presentation is followed by a summary of the discussions that ensued.

FISH, FISHERIES, AND FOOD SECURITY

Cynthia M. Jones, Ph.D.
A.D. and Annye L. Morgan Professor of Sciences,
Eminent Scholar and Director,
Center for Quantitative Fisheries Ecology, Old Dominion University

Cynthia Jones began by outlining the main topics of her presentation—dietary recommendations for seafood consumption, fish production, and the effect of climate change on fish productivity. Jones highlighted the distinction between food security and food safety, noting that food security means having enough food to feed the nation.

Jones noted that eating fish directly is a better option than taking omega-3 fish pills, because, among other factors, someone eating fish may not be eating high-cholesterol foods and, also, fish provides high-quality protein. Typically, the dietary recommendation for fish consumption is a minimum of two servings a week of about 4 ounces, which adds up to about 26 pounds of fish per person per year. To satisfy this dietary recommendation, the American population (approximately 300 million people) would require 7.2 billion pounds of processed fish (i.e., fish sticks, fish fillets, etc.). What amount of whole fish, Jones asked, would be needed to produce that amount of processed fish? She added that her research indicated that there is no consensus on this question. She interviewed the seafood manager at her local market and learned that he was able to extract one-half the weight of a fish in usable fillets—probably more than most people would be able to get, but a usable number for making the calculation. The Food and Agriculture Organization (FAO) of the United Nations also suggests that approximately one-third to one-half of the weight can be extracted depending on the species. With this ratio of one-half processed to unprocessed fish, the answer is that approximately 14.4 billion pounds of whole fish would be needed—though in fact, the average American is eating about 15.8 pounds of fish per year (down from about 16.5 previously), which reduces the total needed harvest, Jones said.

Fish Production

The source of the fish harvest is the commercial and recreational capture fisheries and aquaculture. In passing, Jones pointed out that marine fish are the last commercially captured wild animals in the world—no other wild animal is hunted commercially for food. The U.S. harvest from capture fisheries is mostly fin fish with a small proportion of shellfish. The total of fish landed (i.e., actually brought into port and to market; a certain amount of the catch is discarded at sea) can be assessed in terms of both weight and value (see Table 4-1). Pollock and menhaden top the weight list, at about half of the total. Pollock is made into fish sticks, fillets, surimi, and other fish products. Menhaden (both Gulf and Atlantic) is not considered an edible fish (though it was eaten in the past) but has many other uses, such as animal feed, oil (for human consumption, manufacturing, and fuel), and fertilizer—this type of fish use is called reduction fishery. Salmon and other fish, such as flatfish and cod, make up most of the rest, with shellfish providing the remainder. Looking at the list in terms of monetary values presents a different ranking, however, with crabs at the top followed by salmon, scallops, and lobster.

TABLE 4-1 Major Domestic Species Landed in 2010

Fish	Pounds	Percent	Fish	Dollar Value	Percent
Pollock	1,958,936,000	28.07	Crabs	$572,797,000	16.55
Menhaden	1,471,803,000	21.09	Salmon	$554,816,000	16.03
Salmon	787,740,000	11.29	Scallops	$456,632,000	13.19
Flatfish	625,358,000	8.96	Lobster	$442,735,000	12.79
Cod	557,349,000	7.99	Shrimp	$413,980,000	11.96
Hakes	378,277,000	5.42	Pollock	$291,922,000	8.43
Crabs	349,604,000	5.01	Halibut	$206,553,000	5.97
Squid	337,223,000	4.83	Clams	$200,657,000	5.80
Shrimp	258,972,000	3.71	Cod	$175,060,000	5.06
Herring	253,381,000	3.63	Flatfish	$146,243,000	4.22
Total weight*	6,978,643,000		Total value*	$3,461,395,000	

* This is not the total for all fish landed, only the total for major domestic species landed.
SOURCE: NMFS, 2011.

So, while 14.4 billion pounds are required to fulfill the U.S. dietary recommendation mentioned above, only about 8.2 billion pounds were landed (NMFS, 2011), from which about 1.7 billion went to reduction fishery. Added to that, aquaculture produced about 1.2 billion pounds, and recreational fishing about 200 million pounds—giving a total available poundage of less than 7 billion. In fact, concluded Jones, the nation's needs are not being met by the U.S. harvest.

Turning to the fish harvest in the rest of the world, Jones noted that the worldwide production of fish has risen in the past 50 years, but the marine capture component of that has leveled off since the 1980s, remaining at about 80 million metric tons. Marine aquaculture and freshwater inland fisheries have grown steadily during this time, becoming particularly important in China. Jones felt that there could be an ethical issue with the United States buying fish from countries whose populations might go hungry as a result.

Expanded aquaculture, said Jones, will produce an increased worldwide fish harvest, though at the same time there are problems to be confronted. Such problems include

- habitat destruction: shrimp farming in Asia has destroyed acres of mangrove forests;
- spread of parasites to natural population: sea lice infect many farmed fish and are easily transmitted to the nonfarm fish;
- aquaculture effluent: fish excretion combined with nutrients from excess feed creates pollution of inland and coastal waters;
- genetic hybridization: escaped fish can threaten natural biodiversity; and
- disease transmittal to natural population.

About 60 percent of worldwide fish production is expected to come from aquaculture by 2030, said Jones, with China taking the lead in production.

Climate Change and Productivity

Marine capture and aquaculture produce most of the world harvest, and increases in the harvest are projected to come from increases in aquaculture. However, environmental degradation may affect these projections.

Jones cited a study on climate change effects on marine life which predicted a 25 percent shrinkage in fish size (Cheung et al., 2013). She noted that should this actually occur, there were other results in addition to the obvious one of less edible fish. Reproductive success in fish is proportional to its size: large fish produce more eggs than smaller fish, and therefore more survive to maturity. Thus, the fish populations will shrink along with the size of individual fish, if this prediction is true. The Marine Stewardship Council[1] is predicting that climate change will have a significant impact on fisheries from

- ocean acidification,
- warmer water temperatures, and
- changes in oxygenation levels in the oceans.

---

[1] The Marine Stewardship Council develops standards for sustainable fishing and seafood traceability to increase the availability of certified sustainable seafood. For more information see http://www.msc.org/about-us/what-we-do (accessed September 9, 2013).

Jones provided three examples of the impacts of climate change on fishery productivity, which the Center for Quantitative Fisheries Ecology, which she directs, is studying.

Atlantic Menhaden

Atlantic menhaden is in the herring family, or Clupiedae, and it feeds on plankton. It serves two important functions: first, as prey for our preferred food fish, and second, as a prime ingredient for industrial fisheries. Declining levels have been documented in the fisheries of this very important fish. A new stock assessment shows that they are not overfished at present; overfishing may be occurring, but it has not yet resulted in the population being overfished.[2] Chemical testing in combination with microscopic examination of the fish's earbones can show where they come from and how old they are. Habitat mapping on Atlantic menhaden has found that the upper Chesapeake Bay area, once the most important productive nursery for Atlantic menhaden, has become much less productive in recent years. Jones theorized that habitat loss might be the cause for the declining levels documented in the fisheries.

Atlantic Croaker

Climate change and subsequent warming of water will change the ranges of species, Jones added. The Atlantic croaker is a subtropical species, which used to range from the Gulf Coast to the Chesapeake Bay. However, said Jones, Atlantic croakers are now spawning as far north as New England—quite a large extension in their range. Jones noted that different fish have different sensitivities to temperature change. Atlantic mackerel, for instance, are now being fished off Iceland and thus extending their range; other fish, such as weakfish, may be contracting theirs.

Speckled Trout

Speckled trout is a favorite fish for recreational fishing and for food. Their optimal habitat is seagrass, and the variety present in the Chesapeake Bay—*Zostera marina*, or eelgrass—is at the edge of its southern range and is being replaced by *Ruppia maritima*, or widgeon grass, which is a very ephemeral habitat. Jones questioned the continued survival of this

---

[2] Stock that is recruitment overfished has been brought to the point where it cannot sustain its population. Overfishing may be occurring to a stock, but it is not "overfished" until it is at that point where the population can no longer replenish itself.

fish population should this major nursery habitat disappear from the Chesapeake Bay.

The U.S. dietary suggestions for fish exceed our ability to produce, said Jones, and if U.S. production alone is relied upon, food security for the nation will be compromised.

The impact of changes in management on diminishing the gap between available food and the amount needed would not be significant, she added, emphasizing that, contrary to common perception, the U.S. fisheries are well managed—especially when compared to those in other countries. In the United States only 28 stocks of the total of about 230 are overfished; one in five is being overfished but is not yet formally considered overfished. U.S. stocks are in relatively good health compared to stocks worldwide. This means, added Jones, that substantially increasing productivity through better management is not very likely, though an increase of up to 25 percent has been suggested as possible. However, this may not meet the needs of a growing U.S. population.

Jones admitted that the effects of climate change are unknown. Species are changing their range: some species are doing better while some are doing worse. The timing of events such as spring blooms is changing, which affects the reproductive process and the survival of juveniles: some may not survive to recruit to the adult fisheries if the season is curtailed.

PUTTING THE WORLD ON A FORK

Barton Seaver
Chef and National Geographic Fellow

Barton Seaver explained that his interest in fisheries began with his role as chef, and his deep interest in the ingredients of his craft. In exploring the systems that provided his kitchen, he found himself on a path which led to a Fellowship with the National Geographic, where he works currently, exploring the confluence of ecological and human health.

The strategy of making pronouncements about the importance of oceans has not been effective in influencing popular opinion, Seaver noted. His preferred tactic is to ask people for their opinions, and reflect those back. There are two topics guaranteed to engage everyone's attention: dinner and health—topics very much linked together, and linked to the health of the oceans.

Reading and food were very important to Seaver as a boy, both in themselves and as an introduction to different cultures and opinions. Food, he said, is a powerful tool, particularly as administered through that great bastion of culture, the family dinner table—the communal meal. He quoted a line from John Hersey, "in our quest for food, we begin to find our place in the systems of this world," noting that we are not sovereign over our resources but rather we are *because* of them.

Declining Fisheries

As a boy, Seaver spent his summers by the Patuxent River, fishing, crabbing, and enjoying the amazing bounty of those waters, reeling in blue fish, perch, porgies, croakers, striped bass, and many others. At 25, when he achieved the post of chef he decided to look to his past and serve up the bounty he remembered from his youth. But when he called his supplier, he learned that the crab and fish he remembered were gone—eaten, depleted. It was at that moment that Seaver realized "that the guiding hand of natural selection is quite firmly holding a fork." And he started trying to learn about the systems by which food ended up at his back door and from there onto his guests' plates, and how food so delicious and popular could disappear.

The answer, Seaver discovered, was essentially the tragedy of the commons: the idea that man acting in logical self-interest will ultimately deplete a shared resource, despite understanding that doing so is harmful to everyone's best interests—a narrative he declined to accept. Rather than the story involving human blame and shame, he decided to pursue a different solution. If preventable human error was the cause of this disruption in the ability to feed people, and to maintain jobs and cultural identities, the solution was to prevent that human error—what humans caused, humans could fix. Seaver decided to focus on the impact of the ecosystems on humans rather than the human impact on the ecosystems. This was a more positive dialogue he called the Communion of the Commons: more hopeful and more human that engages people to consider their place in the systems of the world and how to maintain it.

Failure of Current Environmental Campaigns

Many environmental campaigns have traditionally been created in response to acute and isolated issues, Seaver said. They are actionable ideas that engage people to participate in solutions. However, he found that many of these environmental campaigns failed to acknowledge the larger framework of what they were trying to accomplish—to measure the initial cause against the larger context of the action the campaign asked people to engage in.

Citing many examples of this—low carbon footprint, recycling, organics, sustainable farming, fair trade—he noted that initial intent often did not carry through to the ultimate action. For example, organics implied a system by which the earth was not harmed, and profits could be made through products conferring health and wellness on consumers. However, such products as *organic cigarettes* seem at odds with the original intent of the organic movement. And then there is recycling, oftentimes held up as the pinnacle of success in environmental campaigns. And yet for all the reduce/reuse/recycle mantra, it has done very little or nothing at all to dispel the preference for a disposable goods economy—the pace of disposable goods entering the market has actually accelerated. Seaver suggested that many of these environmental campaigns actually end up strengthening current systems and current behaviors in many ways. They often forget to ask the larger question inherent in the concern for the environment: what are we trying to sustain? The answer, said Seaver, is that we are trying to sustain ourselves. Seaver acknowledged that this might be the antithesis of the deep green ideology but the bottom line, Seaver said, is "we're trying to save what we eat, where we live, and the healthful productive lives that nature affords us."

Responsible Policies for Managing Ecosystem Services

Seaver then discussed several instances illustrating the need for the development of responsible policies for managing ecosystem services. As food is a primary driver of human interaction with nature and a principal cause of some of the most detrimental forces that have been visited upon nature, logical policies should be developed to protect ecosystem services that protect human needs (food security) more than human desires (monetary gain), Seaver said.

Food Security Versus Business Profit: Pebble Mine

Seaver turned to the tension of human needs versus human desires through the example of the proposed Pebble Mine in the Bristol Bay area of Alaska. Bristol Bay holds 46 percent of the global sockeye salmon population, and the fishery supports 14,000 jobs. About 7,500 native Alaskans rely on the Bristol Bay salmon for food. The fishery makes $500 million per year and is very successful compared to failing global fisheries. Seaver noted that Bristol Bay fishery is an example of how man can act and live in concert with nature. The proposed Pebble Mine will place a large open pit mine at the very headwaters of the bay. The mine will be lucrative, and will employ many people; it will produce porphyry copper, gold, and molybdenum—valuable minerals with a high demand. There is a great deal of opposition to the mine from fishermen and native Alaskan corporations. Seaver pointed out that the controversy is not about mining per se—the people opposing the mine, for instance, are fishing from metal boats, using GPS systems that require copper wiring and gold circuitry, all dependent on products of the mining industry. The mining industry is crucial to the economy. But, in this case, Seaver believes that the placement of this mine would be a mistake, adding that he saw it as a choice between food security for the human population versus monetary gain. He added that the country supports a strategic oil reserve—perhaps it could also have a strategic food reserve. The very best candidate for such a reserve would be Bristol Bay.

Business Model: McDonald's Corporation Fishery Management

Another solution to managing ecosystem services is the business scenario, with shareholders whose interests are at stake. McDonald's Restaurant Corporation has been very successful in its efforts to restore and manage fisheries, Seaver said. This policy began when one fish supply source after another failed, and the corporation decided that in order keep their restaurants supplied with fish, the best and most economical course would be to make management changes in the fisheries, and thus stabilize their supply source.

Over the past 10 years, McDonald's has followed a global sustainable fisheries program. They have global purchasing standards and perform annual assessments of all suppliers by the Sustainable Fisheries Partnership (SFP).[3] Additionally, 100 percent of McDonald's

---

[3] SFP available at: http://www.sustainablefish.org/about-us (accessed August 14, 2013).

fish worldwide currently comes from Marine Stewardship Council (MSC)-certified fisheries.[4]

The fisheries were saved, and the interests of the shareholders were promoted—a successful and responsible instance of ecosystem services management, said Seaver.

Provide Access Points to the Resources

Seaver emphasized that it is not enough to protect our resources (the "commons")—those who provide these resources also need protection. If there are no farmers there are no farms; if there are no farms, there is no food. Similarly if there are no fishermen, there is no fishery, and without fisheries, there is no seafood. Environmentalism has not come to terms with the idea that human interests and economic systems are vitally important to managing our resources, Seaver said.

Seaver noted that currently farms are subsidized, and a farm is an economic system governed by imperatives of increased production and lowered costs. The success of farms is more and more judged by shareholders. Seaver discussed an alternate model in which farmers are valued by society for what they contribute to their communities, rather than for their quarterly profit statements. When consumers shop at local farmers markets for seasonal items, their needs are met, and the farmer is supported.

Seaver further noted that fisheries are subsidized commodities—fisheries operating over capacity with fleets much larger than needed to catch the allowable quota, thus constraining a very diverse ecosystem into an economic and efficiency model which demands uniformity. Instead of subsidizing the fisheries, Seaver suggested subsidizing the access points to regionalized or direct-from-source products. Fishermen would thus be able to participate diversely in an ecosystem, and to offer to consumers what the oceans give, not to take only from the oceans what the consumer and market wants. Market demand determines the value of products and in this way ensures that only a few species are profitable and thus are landed and brought to market. But all seafood is equally profitable when looked at in terms of fulfilling human needs.

---

[4] Information on McDonald's sustainable sourcing is available at: http://www.aboutmcdonalds.com/mcd/sustainability/library/policies_programs/sustainable_supply_chain/sustainable_fisheries.html (accessed August 14, 2013).

Locally Accessible Versus Imported Food

The current system creates waste and skews toward demands, and not toward sustaining human health, job security, or the ocean's health, Seaver said. This is a market issue and a consumer opportunity as well. The system is forced to create what we want, but this has led to a situation in which about 90 percent of the seafood eaten in the United States is imported, and about 65 percent of the seafood caught in this country is exported, Seaver said. Further, 3 percent or less of what we import is inspected—much imported seafood is mislabeled (fraudulently or erroneously). The result has been a complete decoupling of the product from its source, because we get what we want instead of eating what is accessible to us, Seaver said. Consumers should seek what is available locally rather than request exotic food which is unavailable except by importing food. In this way consumers could act in concert with ecosystems, diversifying demands and focusing on ecosystem services rather than on supply chain services, he said.

Seaver stressed the importance of understanding how we access ecosystem services and how we use them: not just focusing on production models but also the uses and benefits of the resource. He also stressed the need for a behavioral shift between fulfilling our desires and fulfilling our needs. There is no technical solution to any resource management problem that can succeed in absence of a complementary behavioral shift in how we use that product. The solution, Seaver said, is simple: "take what you need, share the rest, and leave the system intact." While the technology of aquaculture and increased crop yields on land may very well give a necessary and vital boost to food production, no alchemy will ever make an "all-you-can-eat buffet" sustainable—neither for human health, nor for the planet. Taking more than one's share and wasting a lot of that is an ongoing human story, Seaver said. Peru, for instance, has the world's largest single-species fishery, landing 10 million metric tons of anchoveta (commonly referred to as anchovy) per year—but 98 percent of that never goes to feed a human being, it all goes into reduction production, much like the menhaden referenced by Cynthia Jones in her presentation. This is not the highest and best use of the product. The 7 billion people of this earth can be fed and their dietary needs fulfilled—but the desires of 7 billion people can never be fulfilled, or even half that many. Needs are finite, but desires are infinite. Environmental sustainability in human health is about better nourishing people with the food available, Seaver said.

The U.S. Department of Agriculture estimates that 14.5 percent of American households are food insecure (Coleman-Jensen et al., 2013). The Centers for Disease Control and Prevention (CDC) says that 35.7 percent of American adults are overweight or obese (Ogden et al., 2012). Current resources are not being used for their highest and best purpose, Seaver said. Seaver emphasized that health is very much linked to food, food sustains life, and people can be no healthier than the foods they consume and how they consume those foods, and the food consumed can be no healthier than the environment that it comes from. Therefore, human beings can be no healthier than the environments that sustain them. If humans desire to lead healthy lives they must equally pursue both wellness and environmental resiliency. This principle does not allow for physical and environmental health to be divorced from each other. The danger in a conversation about ecosystem services is that each is analyzed solo, in isolation, Seaver said.

Human consumption patterns largely define how the world is used and what is consumed defines the how healthy the community is. Seaver encouraged pursuing the resiliency of people and communities and the planet by taking only what is needed, sharing the rest, and leaving the system intact and celebrating wellness over consumption. In the communion of the commons a cultural dialogue about ecosystem services can occur that encourages people, businesses, doctors, and governments to participate in the highest and best use of the commons, which is simply to maximize the profits of our human experience, Seaver said.

AQUACULTURE: ENSURING A FUTURE SEAFOOD SUPPLY FOR A HEALTHY POPULATION AND ENVIRONMENT

Kevan L. Main, Ph.D.
Senior Scientist and Director of Aquaculture Research, Mote Marine Laboratory
President, World Aquaculture Society

Kevan Main began the presentation by highlighting that fish is a primary source of protein for nearly 1 billion people, and global consumption is rising (from 9 kg/person in 1961 to 17.1 kg/person in 2008) (FAO, 2010; OECD/FAO, 2011). Seafood is a high-protein food that is low in calories, total fat, and saturated fat, high in vitamins and minerals, with numerous health benefits. Seafood is the only food source which is still produced from the wild.

However, Main predicted future shortages in food and water will not only constrain growth in terrestrial agriculture but will also require freshwater aquaculture to shift to recirculating systems. The task will be to increase global food security, continue the supply of high-quality seafood, and to restore declining fisheries (i.e., stock enhancements) while limiting the environmental impact on the land, water, and fisheries.

Historically, the oceans were considered limitless—with enough fish to feed the world population. But the ocean harvest has reached the maximum sustainable yield, and the product of capture fisheries has been static for the past 25 years. Aquaculture, on the other hand, is growing and provides an increasingly larger percentage of fishery product every year. The majority of aquacultured seafood is being produced in Asia, with 50 percent from China and another third coming from southeast Asia (Indonesia, the Philippines, and Vietnam)—more than 90 percent of the seafood production occurs in Asia (see Figure 4-1).

The U.S. seafood trade deficit exceeded $10 billion annually in 2011 (NOAA, 2013a). This natural resource trade deficit is second only to oil and gas. Main described this as a food security issue for the United States, which resulted in more than 91 percent of U.S. consumed seafood being imported in 2011 and approximately 50 percent of those imports were farmed (NOAA, 2013a). U.S. aquaculture produced only 5 percent of our seafood supply (NOAA, 2013b); however, seafood safety regulations ensure that U.S.-produced seafood is safe and produced in an environmentally sustainable manner. There is a real opportunity to increase U.S. production to meet the expanding demand for high-quality seafood.

Aquaculture Systems and Environmental Issues

There are a variety of different systems that are used to produce seafood around the world, such as ponds, cages, flow-through systems, and recirculating systems.

Ponds

Ponds are the most common way that seafood is produced around the world. In China, Europe, India, Japan, southeast Asia, and the United States, ponds are used to produce a wide range of different products. The

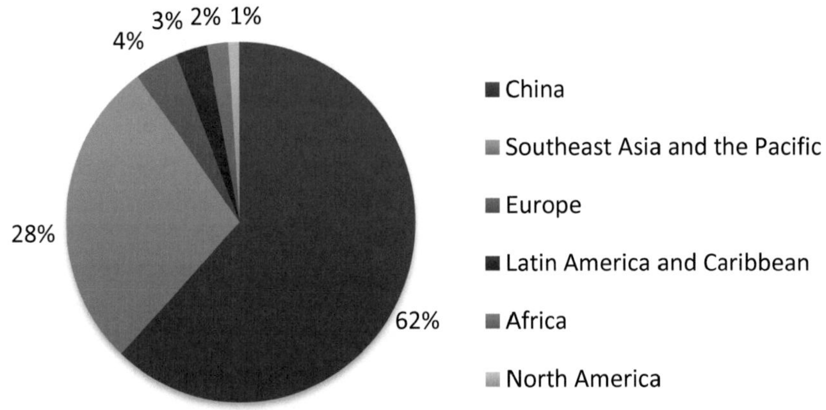

FIGURE 4-1 World fisheries and aquaculture production.
SOURCE: Main, 2012. Data from FAO, 2012.

environmental impact from pond aquaculture can be substantial if the ponds are constructed in critical (i.e., wetland) habitats. Also, if ponds are stocked too heavily and the water is discharged into rivers, lakes, and oceans with poor circulation, there will be detrimental environmental impacts. There is also potential danger for disease to spread to native species, or of nonnative (exotics) species escaping (all aquatic farming systems have these problems).

Near-Shore Cages

The near-shore cage system is the next most common production system. Adverse environmental impact can result if the cages are put in the wrong environment. for example, if the cages are located in a water body where there is poor circulation, if the cages are located too closely together, or if they are stocked too heavily. As with other systems, there is the potential for disease to spread to native species or for exotic species to escape.

Open (Flow-Through) Tank Production Systems

Some of the environmental impacts are reduced when fish are raised inside tanks. However, not only are large volumes of water required for open tank systems, but that water must be discharged, and if concentrated affluent is discharged back into lakes and coastal habitats it can have a negative impact. Escape of exotics and disease spread are also problems in this system.

New Production Systems: Offshore Cage Systems

Offshore cage systems solve some of the near-shore cage problems by moving cage systems out into deep water where the water currents provide better circulation. Offshore aquaculture cages require deep water (a minimum of 30–45-meter water depth). The use of these production systems is expanding around the world (e.g., China, Europe, South and Central America). The growing need to expand food production, and the problems encountered with the coastal systems in particular, have encouraged expansion into offshore cage systems. There has been a lot of technological development in the open-water offshore cages during the past 10 years, involving very sophisticated engineering. These systems can produce large volumes of fish.

Main stressed that it is crucial to ensure these new systems will have limited environmental impact. However, if the cages are placed in the proper locations, and stocked at suitable densities, they will provide large volumes of seafood to meet future demands. She noted that today there are only a few companies that are using offshore cage systems in U.S. waters; it is critical to expand U.S. offshore aquaculture (note that the National Oceanic and Atmospheric Administration is working on the permitting and regulations to allow expansion of these production systems).

Recirculating Aquaculture Systems

Main believes that recirculating land-based aquaculture systems are continuing to grow and this production system is going to play a bigger and bigger role as time goes on. The use of recirculating systems is becoming much more prevalent in European aquaculture and there are a number of facilities that are using the technology in the United States. The disadvantage with recirculating systems is increased production costs, but improved engineering, production of multiple crops, and incorporation of alternative energy can improve the bottom line. Though this technology is relatively new, with the first systems being designed and tested in the 1990s, over the past 10 years there have been a lot of technological improvements in the filtration system components and system designs. In addition, there is expanding support for sustainable environmentally friendly aquaculture and agriculture systems. However, the research needs to focus on reducing production costs and the consumers must be willing to pay more for sustainable products if this technology is to be economically successful.

Main stressed that there are a number of advantages associated with recirculating aquaculture systems, such as conservation of water resources, treatment and recycling of wastewater (thus reducing the environmental impact), and the potential to develop integrated systems (producing both fish and plants). At the Mote Aquaculture Research Park in Sarasota, Florida, there is a commercial farm producing sturgeon and caviar in freshwater recirculating systems, and a marine research unit working on expanding the recirculating systems to produce a wide range of marine species. Figure 4-2 shows a simple schematic for the freshwater recirculating system, in which water moves through a series of filters, the waste is removed using a mechanical and biological filtration processes, and then the water is cleaned, sterilized, and reoxygenated before it is returned to the fish system. The solid waste removed from the water stream can be used as fertilizer for plants. The Mote research unit is demonstrating the production of a wide range of coastal plants, such as mangroves and salt

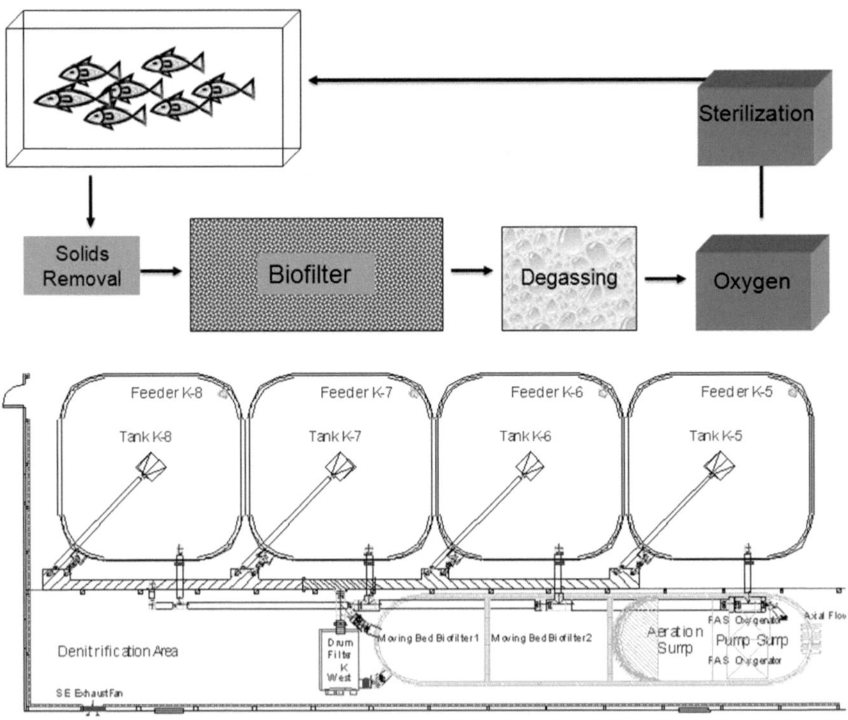

FIGURE 4-2 Recirculating aquaculture system: schematic of a module.
SOURCE: Main, 2012.

marsh grasses, in the wastewater created by marine fish production units—crops that can provide a secondary income stream in addition to that produced by the fish.

Toward Safe and Abundant Seafood

The fact that the United States is importing 91 percent of the seafood consumed and that less than 0.5 percent of those imports are inspected each year highlights a seafood safety issue (NOAA, 2013a). There are a number of seafood concerns to be considered, such as its safety for human consumption, its importance as a source of healthy protein, and the environmental impact on coastal fisheries. For aquaculture, one of the primary issues is the availability and use of high-quality water resources to grow the seafood products.

Water Quality Standards for Recirculating and Other Land-Based Aquaculture Farms

In determining water quality requirements, the water needs to be evaluated at three different points in the system: incoming water, water within the tank or pond, and the water that is discharged from the system back to the environment.

1. **Incoming water quality** *(water entering the animal production system)*. Before the water enters the system it must be evaluated for potential contaminates, to determine if pre-filtration is required. A bioassay where the proposed species (i.e., fish) is grown for a period of time using the incoming water is also recommended.
2. **Water quality in the production system** *(water where the animals are being grown)*. There are no established standards for water quality within the production system, but farming practices are based on knowledge of the tolerance of the species to water quality and water chemistry parameters. The goal is to maintain appropriate water quality and chemistry so that the species will thrive. In the United States drugs and chemotherapeutant use is a major concern and is strictly regulated by the U.S. Department of Agriculture (USDA) and the U.S. Food and Drug Administration (FDA). Use of approved drugs has required withdrawal periods before harvesting. The limited inspection of seafood imports gives even more importance to U.S. production practices. Florida and many other states in the United States have developed best

management practices (BMPs) for aquaculture production. In Florida, the BMPs were designed to preserve the environment and reduce duplications of environmental and other permitting requirements. However, one of the biggest limitations to expanding aquaculture in Florida is limited access and availability of freshwater—despite heavy rainfall, there is an extreme water shortage in Florida, owing to a large population and their needs. How to best allocate this precious resource is a difficult question which state regulators are tackling today.

3. **Effluent water quality** *(water discharged from the animal production system)*. The standards in the United States depend on where the water is being discharged and the volume of animals being grown in the production system. These regulations are set by different government organizations (state agencies for smaller farms or by federal agencies for larger farms) and may require an ongoing water quality and environmental monitoring program.

Safety in Production Processes

The processing of seafood is very strictly regulated in the United States by the USDA and the FDA through a system called Hazard Analysis and Critical Control Points (HACCP). This system was introduced as a preventive approach to food safety as the seafood is processed, rather than trying to detect problems by testing the finished product. Regulations associated with HACCP require frequent monitoring and testing.

Restoring Fishery Resources Through Stock Enhancement

There are several management options for restoring declining fish stocks, such as

- reducing the harvest (with size limits, seasons, closures, trip limits, and so on),
- developing marine protected areas, and
- rebuilding or enhancing the stock through release of cultured animals.

Hatchery-based marine stock enhancement has made impressive research gains since the 1990s, when scientific studies of marine fisheries enhancement were initiated. Following publication of those first scientific studies on the potential to restore marine fisheries resources using stock enhancement in 1989 and the 1990s, more evidence was presented documenting the potential for stock enhancement to restore declining fisheries. Results of small-scale stock enhancement research in Florida indicates that common snook is an excellent model species for achieving and documenting an economically successful stocking program. However, the use of marine stock enhancement requires a "responsible approach" to stocking that is focused on protecting wild fishery resources, Main said.

In closing, Main gave an outline of the important issues she had covered in this presentation:

- Fish is the primary source of protein for nearly 1 billion people.
- Fish is a high-protein food; is low in calories, total fat, and saturated fat; is high in vitamins and minerals; and has been shown to have numerous health benefits.
- Future seafood resources will primarily be supplied by aquaculture.
- The U.S. seafood trade deficit exceeded $10 billion annually in 2011, resulting in a natural resource trade deficit second only to oil and gas.
- About 91 percent of U.S. consumed seafood is imported and about 50 percent of the imports are farmed.
- U.S. aquaculture currently supplies about 5 percent of our seafood supply. It is imperative to increase U.S. aquaculture production.
- Production and safety regulations ensure that U.S.-produced seafood is safe.
- Environmental sustainability is a growing priority in the United States.
- Hatcheries can be a tool to rebuild fisheries if done responsibly.
- Shortages in water and food will constrain growth in terrestrial agriculture (competitive use of land).
- Shortages in water and environmental concerns will continue to shift land-based production to recycle systems.
- Incorporating plant production into recirculating systems has synergistic beneficial effects.

NEW OPPORTUNITIES FOR RESOURCE MANAGEMENT: LIFE-CYCLE ANALYSIS, SUSTAINABILITY, AND CO-BENEFITS

Steven A. Murawski, Ph.D.
Professor and Downtown Progress - Peter R. Betzer Endowed Chair of Biological Oceanography, University of South Florida

Steven Murawski noted that there are different approaches to the notion of general ecosystem services, life-cycle analysis, and its application to fisheries. He explained that his presentation would provide a multidisciplinary perspective on sustainability and ecosystems based on the interactions between people, the water environment, and food supply. His talk would also highlight life-cycle analysis, its operational definitions, its general applications, and its application to fisheries and for addressing sustainability.

Interactions at the Triad

In terms of the people–water environment–food supply triad, Murawski stated, the environment and food production issues are linked to societal well-being and human outcomes as well as general preparedness for environmental catastrophes, including storms, sea-level rise, and harmful algal blooms (see Figure 4-3). The triad presents a sustainability framework to consider what is being extracted from the environment in terms of sustaining people but also in terms of extraction from the feedback loops that develop along the triad. Because there are many varied ecosystem services that people are trying to extract and maintain at the same time, it is important to understand the priority areas for ecosystem services production as well as sustainability.

Life-Cycle Analysis

Murawski explained that life-cycle analysis is an environmental assessment tool to quantify environmental impact throughout the entire life cycle of a product or process. The life cycle of a product includes all phases from raw material extraction, production, transportation, and waste treatment—essentially a cradle-to-grave approach. Life-cycle analysis has a history in industrial process control but has many emerging

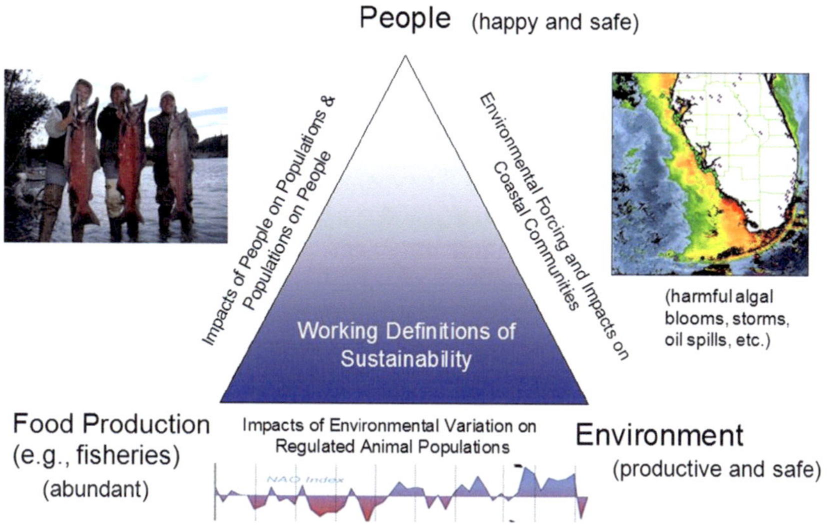

FIGURE 4-3 Interactions at the triad of the people, environment, and food production domains.
SOURCE: Murawski, 2012.

applications relevant to ecosystems. Some of the general applications of life-cycle analysis include (a) evaluation of energy inputs and outputs for biofuel production, (b) transportation system alternatives, (c) greenhouse gas production and evaluation of carbon footprints, and (d) human health outcomes in relation to a variety of issues and settings. When looking at all those life-cycle analyses, Murawski noted, there are differing perspectives on the type of "currency" included in the life-cycle analysis. The currency may be energy, toxic and hazardous substances, food production or consumption, water production or consumption, economics in terms of the consumer or producer, human mortality and well-being, carbon footprint or greenhouse gas issues, or natural resource sustainability.

One example at the nexus of health-related issues and ecosystem services is improving cardiac outcomes for people through changes in dietary consumption. A pooled analysis of prospective studies and randomized clinical trials from Mozaffarian and Rimm (2006) evaluated the relationship between intake of two components of omega-3 fatty acids (eicosapentaenoic acid and docosahexaenoic acid) through fish or fish oil consumption and the resulting relative risk of coronary heart disease death. In this example, stated Murawski, as the intake of the two omega-3 fatty acids increased per week, a break point occurred at 500 milligrams where the relative risk of coronary heart disease death was reduced and remained somewhat constant even with further intake increases (Mozaffarian and Rimm, 2006).

While this is an important assessment, said Murawski, the positive health outcomes need to be balanced against what is referred to as "the seafood dilemma." As consumption of seafood increases, he explained, depending on the individual and the type of seafood, one could be taking an increased load of toxic mercury and other contaminants. The balance between improved health outcomes and actual toxic outcome is important to consider and somewhat controversial. A study from Dickoff and colleagues (2007) looking at balancing the health benefits of reduced coronary heart disease from seafood consumption against the contaminant effects of mercury, dioxins, and polychlorinated biphenyls found the benefits outweigh the risks by 300 to 1 (with a few exceptions). However, achieving recommended levels of seafood consumption (which vary between two 3.5-ounce and two 6-ounce meals per week) is putting increased pressure on the natural resource supply. It appears that there is not enough productivity in the ocean to sustain the required supply both nationally or internationally. This creates conflicting international goals in terms of trying to get people to eat more fish and not having enough fish to eat.

Murawski described another example of applying the life-cycle analysis to fisheries. In looking across all of the dimensions of the fishery process—including fishing, processing, wholesaling, transport, retail, and consumption—a score can be generated in terms of the impact of these various activities on multiple dimensions of sustainability. Results from a life-cycle analysis looking at Danish fish products showed that the fishing stage created the most environmental impacts, mainly due to the energy requirements and partly due to the continued depletion of fish stocks (Thrane, 2006).

Fishery Sustainability

While the world annual seafood production is static among wild sources, the proportion from aquaculture sources is increasing. Murawski proposed that at some point the aquaculture sources will be producing more than wild fisheries. In 2008 the total U.S. fish catch was 8.3 billion pounds with first sale value of $4.4 billion. The U.S. fisheries sales totaled $185 billion in the same year (Van Voorhees and Pritchard, 2009). U.S. imported seafood totaled $28 billion in 2008, of which $13.4

billion was edible seafood and the remainder was used as fish meal in a variety of products (Van Voorhees and Pritchard, 2009).

As a result of the global importance of this issue, there are efforts to rank countries in terms of their compliance with the United Nations Code of Conduct for Responsible Fisheries.[5] While the United States ranks toward the top of overall compliance on a scale of good, pass, or fail, not 1 of the 53 countries landing 96 percent of the global marine catch (based on 1999 values) achieved a score of good (Pitcher et al., 2009). Additionally, looking at Figure 4-4, the United States imports seafood supplies from other countries (indicated with a yellow star in Figure 4-4) with poor records in terms of sustainability scores.

Murawski explained that in the United States fishery sustainability is defined with respect to the size of the stock. For example, overfishing occurs when "the rate of harvest (percent of stock removed by fishing) exceeds the pre-defined maximum rate" (generally about 20 percent per year is sustainable) and overfished occurs when "the current size of the population is less than half of the population size required to generate maximum sustainable yields." In other words, Murawski said, if the rate

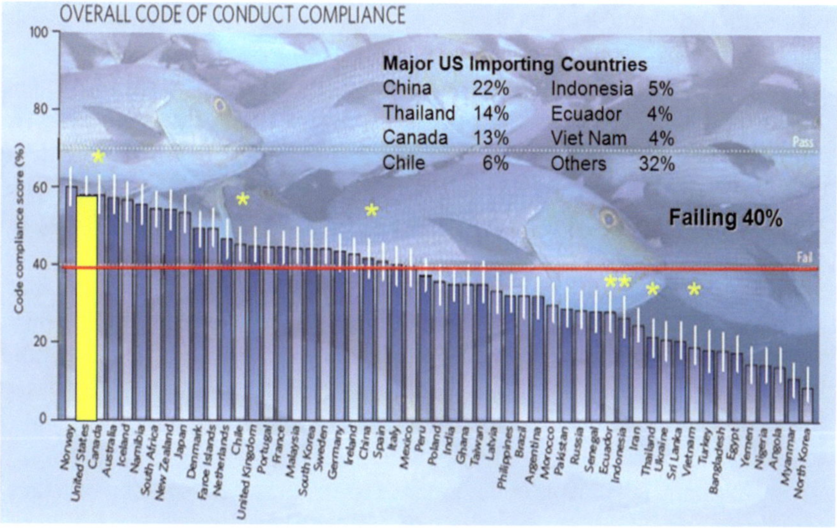

FIGURE 4-4 State of global fisheries.
SOURCE: Murawski, 2012. Data adapted from Pitcher et al., 2009. Reprinted with permission from Nature Publishing Group.

---

[5] United Nations Code of Conduct for Responsible Fisheries is available at: http://www.fao.org/fishery/code/about/en (accessed September 9, 2013).

of outflow exceeds the rate of inflow required to generate sustainable yields, the use rate is not sustainable. Over the past 12 years, there has been steady progress in rebuilding stocks and meeting requirements for sustainable fishery populations in the United States and this progress is projected to continue. Murawski noted that the sea scallop is a dramatic example of rebuilding a stock that was depleted into one that is now sustainable.

Globalization of Fisheries

In looking at the globalization of fisheries, Murawski noted, it is important to balance not only the trade of seafood imports but also the protein deficits that may be created. For example, the European Union has become increasingly dependent on fishery landings from Africa (Swartz et al., 2010), which creates a greater need for grain consumption as an alternative food source in populations that may be protein deficient.

Movements to value sustainably caught seafood and create sustainable certification programs are also occurring at the global level. Life-cycle analysis is being utilized to assess the sustainability of various fishing methods and stocks as a positive feedback to value sustainable fisheries. Murawski noted that this is a regulatory issue for many nations and a difficult one to solve, but the marketplace is showing strong signals to achieving improved sustainability labeling, particularly in Europe. He stated that there are a large number of fisheries trying to achieve sustainable fishing practices and it is hopeful to see the MSC and others providing a certification process to recognize these fisheries in the marketplace.

One of the difficulties to achieving sustainable fisheries, Murawski explained, is the transition cost to society. There are often short-term losses in the life-cycle analysis to achieve long-term gains. In the transition to sustainability, many people will need to change professions, which comes with a major cost. The climate-related aspects of the life-cycle analysis are also very important to consider in terms of greenhouse gas enrichment scenarios and demands on natural ecosystem services, including fish.

Murawski summarized by noting that there is an emerging set of tools to operationalize sustainability and aid in decision making. Life-cycle analysis is one of them and can be done from many different perspectives. The notion of quantifying sustainability is important, as is defining the perspectives that are relevant to the kinds of problems that require multidisciplinary approaches. Finally, it is important to remember that economics is not the only relevant "currency" for ecosystem services.

DISCUSSION

Linda McCauley stated that she was shocked by 91 percent of our fish being imported and 61 percent of domestically produced seafood being exported, and asked the speakers to comment on a local source of fish that may be comparable to locally raised meat in the United States. Main noted that catfish is produced in the southeastern United States and that it is the biggest U.S. food fish production. Additionally, there are smaller-scale producers of tilapia and salmon in the United States and a variety of farms that are trying to develop yellow perch, hybrid striped bass, and other fish. Main stated that the United States is often overly cautious and is not allowing the aquaculture industry to develop and grow. Seaver joined the discussion and noted that it may be best to choose the fish species that farm well, as much of the current aquaculture production in the United States works well in confinement. Unfortunately, he said, we have started down the path of trying to farm what sells well, which is introducing other issues into the system such as genetic modification and negative impacts on ecosystem services.

Christopher Portier from the CDC noted that discussions surrounding climate change, sustainability, and ecosystem services are not even asking questions such as whether people in Atlanta should be eating Dungeness crab, though that is clearly not sustainable. Seaver stated that within the chef community in the United States using frozen fish is still considered taboo even though flash freezing technology has greatly improved. He noted that there are substantial cost benefits to frozen fish in terms of storage, transportation, and consumer usage. Seaver stated that this is an example of how the systematic approaches of sustainability should be acknowledged along with the production aspects.

Lynn Goldman asked Main to comment on the amount of fish feed that is required to go into farm fish production and address whether this is the best way to deliver fish protein. She highlighted that much plant protein is required to develop a pound of beef protein, which has broader impacts throughout the world, and wondered if fish production may have similar global impacts. Main noted that the amount of poundage of fish to feed salmon has decreased and approximately less than one pound of fish is required to grow one pound of salmon. This is a result of research over the past 10 years looking at alternative feeding ingredients and focusing on the alternative protein resources that can be used in fish farming. Main stated that while there have been improvements, this remains an area for further research.

Jack Spengler from Harvard University asked Jones to comment on how well fisheries are managed in the United States and on the associated impacts to communities that depend on the fisheries. Jones noted that some communities are doing less well and established catch limits may move to longer-term targets rather than annual targets. The annual targets tend to result in more drastic highs and lows for the fishermen compared to limits that average over 3 years, for example. Additionally, she stated, there have been economic incentives in the United States (e.g., capitalization funds and low interest rates for vessels) that have built the fishing fleets so large that the available fish catch is divided in an unsustainable way. As a result, Jones said, many fishermen are now part-time, rather than full-time, and many boats sit idle for part of the year, which is a terrible use of capital. Jones noted that one potential solution is to decrease to a sustainable fleet size for the fish catch that is available and start circulating new licenses and catch shares as they come along.

Paul Sandifer from the National Oceanic and Atmospheric Administration noted that in order to continually meet global seafood protein needs into the future, aquaculture will need to be relied on more. The statistics shown here indicate that aquaculture is providing a significant amount of the seafood consumed by Americans and not nearly enough of this is domestic based. Having spent some time in this arena, Sandifer explained that much of the research is pushed by market forces that drive what a producer is allowed to produce.

Sandifer then asked Murawski to comment on whether the United States does indeed take protein away from other places in the world to meet its own demand, and in effect forces a substitution to a less valuable food product (e.g., carbohydrates) in other areas. Murawski responded that there are places where these protein deficits are taking place and the protein source is being converted to export income. The question becomes, what levels of society are benefiting from that export income? He noted that if the generated income does not return to the levels of society where the protein is being removed, then this certainly can be a problem.

Edward Laws asked Murawski to comment on the current aquaculture production of 48 million tons and if this production is sustainable based on a life-cycle analysis. Murawski noted that the vast majority of aquaculture production in the world comes from China; many of the bays and estuaries in China resemble an industrial park for aquaculture. He explained that in utilizing that space for aquaculture other ecosystem services are pushed out (e.g., recreation, nutrient reduction, etc.). Something to also consider is the feed used for the aquaculture products. For instance, Murawski said, if agriculture is used for feed then there may be nonsustainable supply problems. Additionally, the same health benefits (in terms of omega-3 fatty acids) are not available with a feed of only agriculture compared to using fish feed. Murawski went on to say that he does not know if this system will continue to be sustainable in the future.

Bernie Goldstein from the University of Pittsburgh commented on the life-cycle analysis and other sustainability tools presented by Murawski. He asked if some of the economic impacts from fishing and fisheries are included in the life-cycle analysis, as well as further impacts such as the depletion of alternate animal sources in protein-deficient regions and possible emerging infections throughout the world. Goldstein explained that the interagency risk assessment approach within the U.S. government facilitated an opportunity to incorporate issues that were on the borders into the assessments and asked if similar harmonizing efforts could be utilized with life-cycle analysis. Murawski noted that starting with basic metrics, it would be helpful to include these diverse societal goals into the life-cycle analysis without it becoming unwieldy. The classic fishery metrics are how big the stock is, and how fast is it leaving. However, the transition costs bring in a variety of other metrics. In the United States, Murawski said, not only is preventing overfishing a metric but there are nine other national standards that target economic efficiency, balance between the states, impact on coastal communities, and others that layer different dimensions into the analysis. Murawski stated that it will be interesting to see if commonality of metrics develops at the global level. Additionally, one could think about using development tools or a broader set of nonregulatory controls to address income disparities and other impacts on the developing world as incentives to incorporate sustainable practices into fisheries management.

REFERENCES

Cheung, W. W. L., J. L. Sarmiento, J. Dunne, T. L. Frölicher, V. W. Y. Lam, M. L. Deng Palomares, R. Watson, and D. Pauly. 2013. Shrinking of fishes exacerbates impacts of global ocean changes on marine ecosystems. *Nature Climate Change* 3:254–258.

Coleman-Jensen, A., M. Nord, and A. Singh. 2013. *Household food security in the United States in 2012*. ERR-155. Washington, DC: U.S. Department of Agriculture, Economic Research Service.

Dickoff, W. W., T. K. Collier, and U. Varanasi. 2007. The seafood "dilemma": A way forward. *Fisheries* 32(5):244–246.

FAO (Food and Agriculture Organization of the United Nations). 2010. *The state of world fisheries and aquaculture 2010*. Rome: FAO.

FAO. 2012. *The state of world fisheries and aquaculture 2012*. Rome: FAO.

Main, K. 2012. *Aquaculture: Ensuring a future seafood supply for a healthy population and environment.* Presentation at the Institute of Medicine workshop on Understanding the Connections Between Coastal Waters and Ocean Ecosystem Services and Human Health: Basic Services, Valuation and Resiliency. Washington, DC.

Mozaffarian, D., and E. B. Rimm. 2006. Fish intake, contaminants, and human health: Evaluating the risks and the benefits. *Journal of the American Medical Association* 298(15):1885–1899.

Murawski, S. 2012. *New opportunities for resource management: Life-cycle analysis, sustainability, and co-benefits.* Presentation at the Institute of Medicine workshop on Understanding the Connections Between Coastal Waters and Ocean Ecosystem Services and Human Health: Basic Services, Valuation and Resiliency. Washington, DC.

NMFS (National Marine Fisheries Service). 2011. *Fisheries of the United States, 2010*. Silver Spring, MD: NMFS.

NOAA (National Oceanic and Atmospheric Administration). 2013a. *FishWatch: Outside the U.S.* Available at: http://www.fishwatch.gov/farmed_seafood/outside_the_us.htm (accessed August 14, 2013).

NOAA. 2013b. *State of the coast: Marine aquaculture—much potential, many challenges*. Available at: http://stateofthecoast.noaa.gov/com_fishing/aquaculture.html (accessed August 14, 2013).

OECD/FAO (Organisation for Economic Co-operation and Development/Food and Agriculture Organization of the United Nations). 2011. *OECD-FAO agricultural outlook 2011-2020*. Available at: http://www.oecd.org/site/oecd-faoagriculturaloutlook/48184313.pdf (accessed August 14, 2013).

Ogden, C. L., M. D. Carroll, B. K. Kit, and K. M. Flegal. 2012. *Prevalence of obesity in the United States, 2009–2010.* NCHS Data Brief No. 82. Washington, DC: National Center for Health Statistics, Centers for Disease Control and Prevention.

Pitcher, T., D. Kalikoski, G. Pramod, and K. Short. 2009. Not honouring the code. *Nature* 457(5):658–659.

Swartz, W., U. Rashid Sumaila, R. A. Watson, and D. Pauly. 2010. Sourcing seafood for the three major markets: The EU, Japan, and the USA. *Marine Policy* 34(6):1366–1373.

Thrane, M. 2006. LCA of Danish fish products: New methods and insights. *International Journal of Life Cycle Assessment* 11(1):66–74.

Van Voorhees, D., and E. S. Pritchard. 2009. *Fisheries of the United States, 2008.* Current Fishery Statistics No. 2008. Silver Spring, MD: National Marine Fisheries Service, Office of Science and Technology.

5

Opportunities for Medicines

The marine ecosystem has been a tremendous resource of novel chemistries, many of which have successfully been translated into or have inspired new medicines. This chapter presents an overview of products that have originated from the marine environment. The presenter cautions that future development of pharmaceutical products may be compromised by ecosystem stressors such as overfishing and selective fishing, and increased ocean acidification and nitrification.

CURRENT AND EMERGING OPPORTUNITIES AND CHALLENGES FOR PHARMACEUTICALS FROM THE SEA

William H. Gerwick, Ph.D.
Scripps Institution of Oceanography and
Skaggs School of Pharmacy and Pharmaceutical Sciences,
University of California, San Diego

William H. Gerwick began by noting that the discovery and production of medicines and other natural products is an often overlooked but highly critical ecosystem service and is closely linked to the biodiversity function discussed elsewhere in the workshop. He then outlined the main questions in his discussion about the search for pharmaceuticals from the sea:

- Why are new pharmaceuticals needed?
- Are oceans a productive source of new medicines?
- Are there examples of research and drug discovery from the sea?
- Will environmental stressors affect drug discovery from the sea?

The Need for New Pharmaceuticals

Cancer is one of a number of diseases that are ineffectively treated by current medications. Although we see reductions in breast, bronchial, colorectal, lung, stomach, and uterine cancer in women—due to modifications to behavior like smoking, refrigeration of foods, and early screening—there are other cancers that are increasing over time, and currently cancer is the cause of one in four deaths for both men and women (Siegel et al., 2012).

There are many other types of diseases that are poorly treated, such as microbial infections. *Staphylococcus aureus*, for example, has gone from being a highly sensitive strain to an increasingly drug-resistant strain (methicillin-resistant *S. aureus* that is now encountered in the community as well as in hospitals). It is a danger to population health, and current pharmaceuticals do not work effectively against the infection (Chambers and DeLeo, 2009). New pharmaceuticals are needed to treat these types of resistant organisms, Gerwick said. New methods for their application are needed as well because the same problem will recur if antibiotics continue to be used in the way they have in the past. Also newly emergent diseases, particularly viral diseases, are being transmitted from wild animal populations to human populations and giving rise to various diseases such as AIDS and many others. As humans continue to erode the natural habitat, there will be more contact with wild animals and an increased risk of viral diseases in the wild animal population transmitted to human populations.

Pharmaceuticals are derived from diverse sources, Gerwick explained. Of 1,355 new approved drugs spanning the period 1981 to 2010, about 26 percent were derived from natural products. A growing number of pharmaceuticals, about 21 percent, come from biologics or vaccines. Just over half of pharmaceuticals are synthetic drugs, but it can be seen that in many cases the synthetic chemist has looked to nature for an aspect of a molecule and then embedded this special feature into another molecule of synthetic origin, which now has the needed pharmaceutical properties. In this way, a majority of pharmaceuticals are in some sense natural product derivatives or inspired by natural products (Newman and Cragg, 2012).

The Oceans Are a Productive Source of New Medicines

The marine environment and its unique life forms, with their myriad colors and shapes and sizes and adaptations to underwater life, have been a tremendous resource of novel chemistries, many of which have successfully been translated into new medicines. Gerwick mentioned nine marine natural products, derivatives, or inspired agents approved by the U.S. Food and Drug Administration (FDA) or European Medicines Agency (EMEA) (see Figure 5-1). For example, there is a series of pure nucleosides deriving from sponges which has given rise to three very useful anticancer agents and one antiviral agent. Then, there is a peptide used by the cone snail to prey on fish, which has been used in medications for treatment of chronic pain that is no longer responsive to opioids. These and many other new drugs look to nature for inspiration. Gerwick pointed out that the success record of marine natural products in this field, namely one drug per 2,450 compounds, is six times more productive than the industry standard (Gerwick and Moore, 2012).

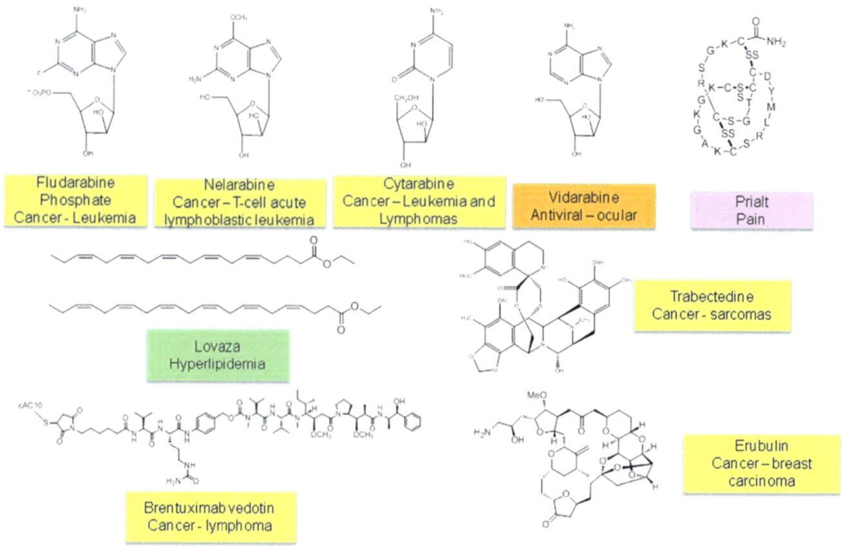

FIGURE 5-1 Nine marine natural products, derivatives, or inspired agents approved by the U.S. Food and Drug Administration or European Medicines Agency.
SOURCE: Gerwick and Moore, 2012. Reprinted with permission from Elsevier.

Examples of Current Research and Drug Discovery Process

Gerwick then turned to one of the current subjects of study at his laboratory—marine cyanobacteria, which are very ancient and very abundant organisms, very rich in structurally diverse bioactive natural products, with a lot of genetic capacity, making them very useful to the laboratory. He stressed that not only is the chemistry of the organism studied and preserved, but also the DNA—science is increasingly moving into the genetics of natural products biosynthesis. Gerwick noted that the process of drug discovery from marine algae and cyanobacteria supports multiple collaborations—in the private sector, in the government, in universities, and internationally. Many important new drugs will be developed through the study of the biosynthetic process.

The areas of highest biodiversity—on both land and sea—are concentrated on the tropic belt, where plants and animals are all competing to harvest sunlight and to access nutrients. This very competitive environment seems to encourage the production of biologically active molecules. Gerwick described the collection of cyanobacteria and other organisms from the mangrove swamps in Curaçao, and remarked on the danger of infectious disease the researchers encountered in these places, but stressed that important drugs have resulted from these efforts, including a very powerful anticancer drug.

Gerwick described the technical aspects of the biosynthetic processes that go into discovering new natural product structures that are useful in the development of drugs. The team took a cancer cell line assay to guide the isolation of a couple of molecules and, by using their standard techniques, came up with some novel structures, which had exactly the same configuration of epoxy ketones as a class of drug just entered into the market—a very powerful anticancer drug. Then, having created a chemical synthesis for this compound, they were able to use that process to make about 12 additional analogue structures. In studying these molecules, they learned that some had very high potency at killing cancer cells and some were inactive, which demonstrated structure activity relationships in this drug class.

Gerwick continued onto a different area of science involving the study of how the compound binds with an enzyme. This particular compound binds in a way unlike any of the other drugs in this class, revealing new ways to interact with the human proteasome, which is a very effective drug target. The human proteasome is a protein degradation "machine" in cells that breaks down proteins. It is a very good target for anticancer therapy, and a target for many other therapies. There is a possible use in neurological disorders, said Gerwick, describing an experiment using very low concentrations of the analogue under study, which caused neurons to develop neurite projections, becoming highly branched. This could offer a therapy for various conditions such as stroke and spinal cord injury. Gerwick suggested that this molecule class has promise for both cancer and the neurosciences.

Gerwick then described expeditions in Papua, New Guinea, harvesting cyanobacteria. The Papuan samples exhibit an exotic-looking cyclic molecule, which has some very exceptional cytotoxic activities, with profound cancer cell toxicity. An experiment administering the compound in cancer research is showing much promise. The structure of the molecule has inspired a great deal of interest, Gerwick said, and his research team is conducting further study on the subject. He described the method for deriving pure cultures of the bacteria from environmental samples. They were eventually successful in isolating the cyanobacteria, and then proceeded to sequence much of the genome of the organism, finally succeeding in describing the entire biosynthetic gene cluster (see Figure 5-2). The process is an astonishing example of modern technology, which has captured the imagination of many—this might be a way of harnessing the power of the cell; a pathway toward the manufacture of new molecules, Gerwick said.

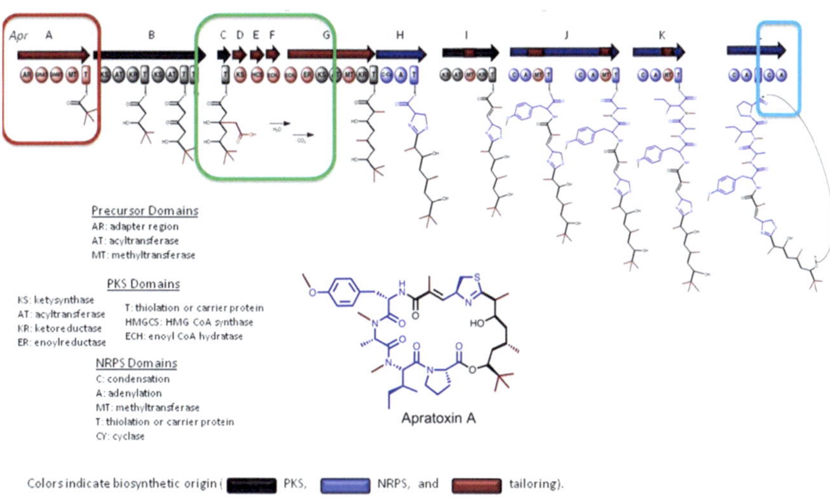

FIGURE 5-2 Apratoxin A cluster annotation and proposed product assembly.
SOURCE: Grindberg et al., 2011. Reprinted with permission.

How Will Environmental Stressors Affect Our Ability to Discover New Pharmaceuticals from the Sea?

Gerwick then moved to environmental stressors and their impact on the discovery of new pharmaceuticals from the sea. He noted several issues, including

- overfishing,
- selective fishing (removal of apex predators),
- nutrient inputs,
- climate change in ocean, and
- ocean acidification.

Human activities have dramatically altered species composition in tropical reefs, said Gerwick, taking as an example the Northern Line Islands (a chain of atolls and islands in the Pacific). The islands get progressively more and more inhabited, going from the uninhabited Kingman reef south to Christmas Island (population of 5,500+). In the north islands, the researchers were surprised to find that the majority of the biomass consisted of large predators, such as snappers, sharks, turtles, and dolphins. Gerwick noted that the expected food pyramid—large amounts of small species on the bottom supporting small amounts of large predators on the top—was turned upside down in this instance. The majority of the biomass in these uninhabited areas was in the large predators: there was little algal biomass, a few herbivorous fish, a few more of the small carnivores, but most of the biomass was resident in the large predators. However, as one goes down toward Christmas Island, the number of large predators grows less and less.

There are substantial inputs of nutrients into the ocean from human population at Christmas Island, which correlates directly with increasing cover of algae and decreasing coral cover, leading to a very degraded environment at Christmas Island. However, Gerwick noted that, ironically, this degraded environment supports a rich complement of cyanobacteria—the source of the inspirational molecules that he had discussed earlier (Sandin et al., 2008).

Gerwick next touched on the effect of ocean acidification (a product of climate change) on natural products, using the example of an experiment studying the production of secondary metabolites in cyanobacteria under different levels of acidification. They hypothesized that ocean acidification may modulate the growth and toxicity, and that such alterations could have substantial environmental impacts, as well as affect drug discovery efforts. They found that as the pH levels became more acidic, there was a destruction of the cellular structures, a decrease in biomass, and a substantial decrease in the production of the natural products they were tracking. Gerwick stressed that this was a small study, and incomplete, but that a preliminary conclusion might be that ocean acidification may decrease growth in natural products production from marine cyanobacteria, thereby decreasing their competitiveness, as well as opportunities for drug discovery.

Gerwick closed his presentation by opining that, both historically and currently, nature is the best source of inspiration for the discovery of new pharmaceutical agents, and that turning away from this caused some significant problems in the pharmaceutical pipeline, now remedied by a return to natural products. Further, he stated that the integration of innovative methods in natural products chemistry is enhancing and accelerating our exploration of the chemical and biosynthetic capacities of marine life, and that this is a vigorous and multidisciplinary field which holds great promise.

He noted that the marine environment is rapidly changing due to human activities. Further, we have only a fragmentary knowledge of human impacts on the adaptations of marine life to their habitats and, ultimately, their success as species. Our health, he said, as realized through new pharmaceuticals from the sea, is intimately tied to ocean health, and it is in our best interest to maintain it.

REFERENCES

Chambers, H. F., and F. R. DeLeo. 2009. Waves of resistance: *Staphylococcus aureus* in the antibiotic era. *Nature Reviews Microbiology* 7(9):629–641.

Gerwick, W. H., and B. H. Moore. 2012. Lessons from the past and charting the future of marine natural products drug discovery and chemical biology. *Chemistry & Biology* 19(1):85–98.

Grindberg, R. V., T. Ishoey, D. Brinza, E. Esquenazi, R. C. Coates, W. Liu, L. Gerwick, P. C. Dorrestein, P. Pevzner, R. Lasken, and W. H. Gerwick. 2011. Single cell genome amplification accelerates identification of the apratoxin biosynthetic pathway from a complex microbial assemblage. *PLoS ONE* 6(4):e18565, doi:10.1371/journal.pone.0018565.

Newman, D. J., and G. M. Cragg. 2012. Natural products as sources of new drugs over the 30 years from 1981 to 2010. *Journal of Natural Products* 75(3):311–335.

Sandin, S. A., J. E. Smith, E. E. DeMartini, E. A. Dinsdale, S. D. Donner, A. M. Friedlander, T. Konotchick, M. Malay, J. E. Maragos, D. Obura, O. Pantos, G. Paulay, M. Richie, F. Rohwer, R. E. Schroeder, S. Walsh, J. B. C. Jackson, N. Knowlton, and E. Sala. 2008. Baselines and degradation of coral reefs in the Northern Line Islands. *PLoS ONE* 3(2):e1548, doi:10.1371/journal.pone.0001548.

Siegel, R., D. Naishadham, and A. Jemal. 2012. Cancer statistics, 2012. *CA: A Cancer Journal for Clinicians* 62(1):10–29.

6

Coastal Change and Human Health

This chapter provides a summary of a presentation on the role of coastal ecosystems in protecting Gulf Coast communities; the Louisiana Gulf Coast is discussed as an example. The presentation highlights the role that intact coastal ecosystems (barrier islands and wetlands) can play in mitigating flooding and in wave attenuation. These ecosystems play a critical role in mitigating low-magnitude, high-frequency events such as storm surges and flooding that can be disruptive to community life. The summary of the presentation is followed by a summary of the discussion that ensued.

THE ROLE OF COASTAL ECOSYSTEMS IN PROTECTING GULF COAST COMMUNITIES

Denise J. Reed, Ph.D.
Chief Scientist,
The Water Institute of the Gulf

Denise J. Reed began by stating that her presentation would focus on ecosystem services and coastal protection and the notion that natural systems can protect coastal communities. She went on to explain that coastal areas are dynamic places. They are modified by the changes people make to natural systems, for example, the effect of increased population pressure on ecosystems. Some natural systems remain intact in isolated places in Alaska which experience less population pressure and important ecosystems continue to thrive in those areas. Much can be learned by contrasting modified ecosystems and those that remain in a natural state to determine an appropriate point between the two extremes that could be aimed for, she said.

Louisiana Coastal Ecosystem

Reed mentioned that as a Louisiana resident she has directly observed changes in the coastal systems in Louisiana. She noted that since the 1930s, coastal Louisiana has lost 1,880 square miles from land to open water. Currently the coast is losing more than 16 square miles per year. It is estimated that the coast may lose up to an additional 1,750 square miles of land over the next 50 years without any increase in coastal preservation.

In Louisiana's Comprehensive Master Plan for a Sustainable Coast,[1] an estimate was developed for the impact of a 100-year storm 50 years into the future. This scenario is for a storm that has a 1-in-100 chance of occurring in any given year. Under this scenario, the whole coast of Louisiana would be in the floodplain (0–5 feet) and some communities would be very deep in the floodplain (20–25 feet). The potential damages of such a storm would reach up to $23.4 billion annually and there would be implications for people's lives, jobs, communities, and the economy (Coastal Protection and Restoration Authority of Louisiana, 2012).

The Protective Role of Coastal Ecosystems

Reed then posed the question, what if we could reverse land loss trends? An analysis included in the master plan showed that if $25 billion dollars were spent on the best projects, land lost could be curtailed, not returned, but land lost could be stopped or reach equilibrium. What would that do for flooding for coastal communities? What would it mean for coastal communities? What would it mean for the culture of the communities and the lifelines or supporting infrastructure for communities (roads, power lines, and other services and structures essential for sustaining life)? In response to these questions, Reed stated that science provides some answers.

Reed discussed a study by Leonard and Croft (2006) that demonstrated that wetlands do change how water flows through them. In this study the water flow velocity, turbulence, and total turbulent kinetic energy were different in land bodies with a *Spartina alterniflora* marsh canopy compared to those without. Studies also indicate that the coastal

[1] Louisiana's Comprehensive Master Plan for a Sustainable Coast is available at: http://issuu.com/coastalmasterplan/docs/coastal_master_plan-v2?e=3722998/2447530 (accessed August 14, 2013).

plant community can affect the way water moves through the wetlands. Plants have individual properties and community properties that affect wave dissipation (Anderson et al., 2011). Seaweed fronds, for example, recline against each other to withstand the high velocity of waves and mangrove density moderates the role of wave breaking on overall wave attenuation. A meta-analysis of studies on the protective role of coastal marshes (Shepard et al., 2011) found that salt marshes or coastal wetlands that are flooded and drained by saltwater brought in by the tides (such as those found along the Gulf Coast) have value for coastal hazard mitigation and climate change adaptation. Factors identified in potentially influencing wave attenuation included marsh species identity or type, density, height, percent cover, and patch size and location. Factors that correlated with shoreline stabilization included species identity, vegetation density, vegetation height, and biomass production.

Reed went on to discuss storm surges, an issue of great public concern, and the role of wetlands to mitigate their impacts. He described a study conducted by Loder and colleagues (2009) that employed numerical modeling to determine the effect of the elevation of the marsh (high or low) and storm surge elevation. They found that higher marsh elevation has a protective impact on storm surge elevation. The lower the marsh, the lower the protective effect on storm surge and as the storm surge rises, the protective effect of the marsh is eliminated. Reed said these findings point to where and when this protective ecosystem service can be most effective. The study also highlights the variability in the potential protective effects in ecosystem services because storm surges and marsh elevation are not static.

The variability in protective effects of ecosystem services has also been explored in mangroves and wave attenuation, Reed said. Studies have found that wave attenuation is influenced by a number of factors, including the species of mangrove, their location, and age of trees. Studies also indicate that the protective effects of mangroves on wave attenuation are very variable.

Ecosystem Restoration and Mitigation of Low-Magnitude, High-Frequency Events

Reed went on to discuss how ecosystem restoration can assist coastal communities. She noted that there are many low-magnitude, high-frequency events that affect everyday life that need to be considered and managed. Resilient coastal ecosystems are an ecosystem service that could play a role in mitigating these events. Big flood events, she said, will inflict damage even with an "intact coast" and for these, communities must prepare, adapt, and at times evacuate. She also noted that sea-level rise increases flooding over time even without coastal change; this situation will have to be addressed.

Reed emphasized that messaging should not ignore the lower-magnitude, higher-frequency nondisaster events where actions to restore coastal habitats can make the most difference. Road floods are an example of this type of event. When roads flood, school bus and other transportation is disrupted making it a challenge to get children to school and employees to work. Flooding also affects yards and parks and can reduce recreational opportunities. Houses may need to be elevated in flood areas, which can become a significant problem for elderly and disabled individuals who have difficulty navigating stairs. Flood areas are associated with increased mosquitoes and the public health problems they pose. In some rural communities that rely on septic tanks, these do not work when water levels are too high and can pose important sanitation problems.

In closing, Reed stated that the important message for coastal protection is the way in which intact, restored barrier islands, and wetlands interact with coastal water movements. It is not about big events such as Katrina, but the protection ecosystem services provide in mitigating the moderate events that happen more frequently. Further, the additional services (recreation, fisheries) and value provided by ecosystems also provide another reason to focus on the natural systems that can protect coastal communities.

DISCUSSION

Jay Lemery began the discussion by asking Reed about the mindset for preventive action in the Gulf, especially after Katrina. Reed responded that the mindset in the Gulf is "you pay now or you pay later," and it is getting more and more difficult to expect people living in the area to start over again.

An audience member commented that humans do things to the ecosystem and expect nature to behave differently—for example we build in low-lying areas and expect that the surge will not affect the area. What needs to change is human nature, not nature. Humans need more common sense in how these ecosystems are used.

Reed responded that we need to make tough decisions regarding how to invest in coast-dependent activities. There are some activities which need to be located near the water such as port activities. Oil and refinery facilities also need to be near water and on flat land; the financial sector on the other hand does not need to be located near water. We need to be more strategic with undertaking water-dependent operations and securing them.

Christopher Portier and Linda McCauly both commented on the adverse impact of high-frequency events and ecosystem degradation on communities along the coast. Reed gave an example of a Native American community that has been dwindling over time in response to frequent events and ecosystem degradation. She said that the chief of the community has made a push to relocate the people as a group together to another place where they can still have a sense of community. Reed said that people tend to move individually, leaving the community unviable; ultimately there is a loss of social cohesion and the sense of community is lost.

Linda McCauley referred to one of Reed's slides showing the extent of land that would be underwater and not sustainable. Reed commented that a plan for the future for such areas is needed, but must have reasonable expectations. Building up marshlands or viable wetlands can make a difference but they can only do so much. She reiterated that in the idealized analysis she discussed earlier (that would spend $250 billion over 50 years) would result in a different future but it would never return the ecosystem to its previous state.

Reed emphasized that from a ecosystem protection perspective, it is important to communicate to the public about those high-frequency events because people can relate to them and remember them. These events also provide a basis for further communication about more complex choices and longer-term decisions at the household and community level.

REFERENCES

Anderson, M. E., J. McKee Smith, and S. Kyle McKay. 2011. *Wave dissipation by vegetation*. ERDC/CHL CHETN-I-82. Vicksburg, MS: U.S. Army Engineer Research and Development Center.

Coastal Protection and Restoration Authority of Louisiana. 2012. *Louisiana's comprehensive master plan for a sustainable coast*. Baton Rouge, LA: Coastal Protection and Restoration Authority of Louisiana.

Leonard, L. A., and A. L. Croft. 2006. The effect of standing biomass on flow velocity and turbulence in *Spartina alterniflora* canopies. *Estuarine, Coastal, and Shelf Science* 69(3–4):325–336.

Loder, N. M., M. A. Cialone, J. L. Irish, and T. V. Wamsley. 2009. *Idealized marsh simulations: Sensitivity of storm surge elevation to seabed elevation.* ERDC/CHL CHETN-I-78. Vicksburg, MS: U.S. Army Engineer Research and Development Center.

Shepard, C. C., C. M. Crain, and M. W. Beck. 2011. The protective role of coastal marshes: A systematic review and meta-analysis. *PLoS ONE* 6(11): e27374, doi:10.1371/journal.pone.0027374.

7

Ensuring Benefits of Recreational Waters Through Monitoring

This chapter focuses on one of the ecosystem services provided by natural water—recreational opportunities. Systematic monitoring of recreational waters is key to ensuring that people can benefit from this ecosystem service. A summary of some of the challenges of water quality monitoring is presented and is followed by a summary of the discussion that ensued.

USE OF INDICATOR ORGANISMS TO ASSESS PUBLIC HEALTH BENEFITS AND RISKS ASSOCIATED WITH RECREATIONAL USE OF NATURAL WATERS

Edward Laws, Ph.D.
Professor, Louisiana State University School of the Coast and Environment

Edward Laws began his presentation by highlighting that the monitoring of recreational waters for the purpose of deciding safe use for the public is unfortunately in a very imperfect state of affairs. Generally, monitoring of recreational waters is a responsibility of local health departments, although the U.S. Environmental Protection Agency (EPA) is charged with setting up standards, confirming test results, and facilitating labs and technology for localities. Laws noted that there are a number of issues surrounding recreational use of natural waters, some of which are close to being resolved and others that remain unclear, which will be presented in his talk.

Valuing Recreational Waters

Laws discussed a paper by Costanza and colleagues (1997) in which the value of ecosystem services, including recreational waters, was ranked and illustrated according to natural capital. The authors described the intrinsic connections between human activities and the environment and argued that there is a benefit to human welfare added to the market values of these services (Costanza et al., 1997). Laws highlighted a list of ecosystem services and the value, in billions of dollars per year, given in the analysis (see Box 7-1). Nutrient cycling was perceived as the biggest ecosystem service provided, and farther down the list recreation was found in ninth place, worth $815 billion per year. It is interesting to note that most of these ecosystem services are aquatic.

Laws noted that Kamehameha Schools in Hawaii initiated a program for students as part of the Center for Ocean and Human Health at the University of Hawaii. The program is for children of Hawaiian ancestry and has played a role in motivating students in the conservation of recreational waters and preservation of their cultural practices. In this example, the value placed on recreational waters in Hawaii was leveraged to support education and conservation.

Monitoring of Recreational Waters

In 1972, the EPA began to address the monitoring of pathogens in recreational waters. Laws highlighted that at the time there was remarkably little information on which to base water quality criteria. There were

BOX 7-1
Value of Ecosystem Services (billions of dollars per year)

Nutrient cycling	$17,075
Cultural	$3,015
Waste treatment	$2,277
Disturbance regulation	$1,779
Water supply	$1,692
Food production	$1,386
Gas regulation	$1,341
Water regulation	$1,115

SOURCE: Costanza et al., 1997.

some studies on concentrations of fecal coliform bacteria and sampling for salmonella. The results of those studies indicated that if the samples contained between 1 and 200 fecal coliforms per hundred milliliters, then 32 percent of the samples likely contained salmonella. As the number of fecal coliform counts increased, the percentage of the water samples that contained salmonella also increased; for instance, more than 2,000 fecal coliform measured per hundred milliliters indicated that 97.6 percent of the water samples likely contained salmonella.

Laws noted that the EPA then organized epidemiological studies to improve the quality of evidence. For the studies, the EPA needed to find recreational waters in the United States that were contaminated with fecal pollution and where people were utilizing the waters for swimming. The EPA identified both marine water and freshwater that were assayed for a list of indicator organisms that would be associated with human feces and then correlated those data with the incidence of people getting sick from going in the waters (defined as submerging one's head under water). Laws highlighted that with these EPA studies the number of times the person went under water or the amount of water the person swallowed did not make a difference; it was all counted the same. In comparison, if the studies were conducted in the United Kingdom, the amount of time a person was in the water and activities they participated in would have been assessed. Roughly 7 to 10 days after people were identified as going in the water the EPA would follow up with these individuals and ask if they had gotten sick. Reported illnesses associated with ingesting pathogens related to feces were recorded as highly credible gastrointestinal illnesses and were correlated with the previously assessed indicator organisms.

Laws explained that *Enterococcus* and *Escherichia coli* (*E. coli*) bacteria are found in everyone's gastrointestinal system and are shed with fecal matter (regardless of whether people are healthy or sick); as such, if feces were present in the water, one would expect to find these bacteria. Figure 7-1 shows the statistical regressions that were developed for marine water in which the correlation for *Enterococcus* is statistically significant but the correlation for *E. coli* is not and is much more scattered. The statistical regressions for freshwater are not displayed, but Laws noted that both *Enterococcus* and *E. coli* correlations were statistically significant in freshwater samples.

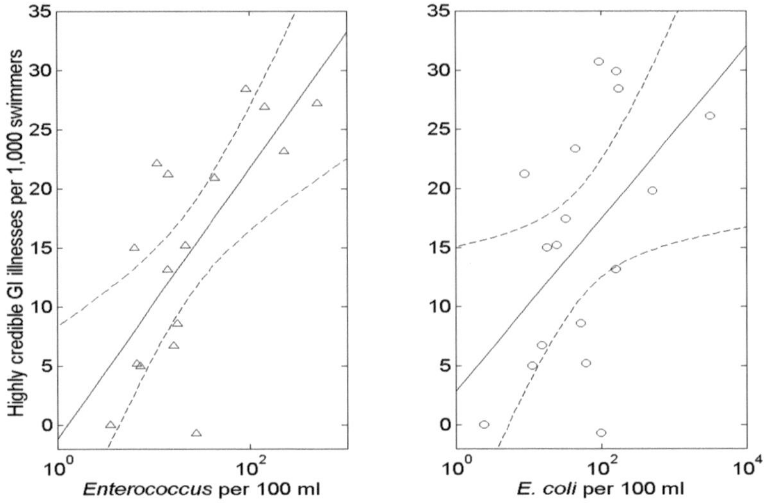

FIGURE 7-1 Marine water correlation between highly credible gastrointestinal illness in swimmers and bacterial concentrations.
NOTE: The y axis represents the occurrence of highly credible gastrointestinal illness per 1,000 swimmers and the x axis represents the concentration of bacteria in the water per 100 mL based on a logarithmic scale.
SOURCE: Cabelli, 1983.

The EPA released this marine and freshwater criterion for fecal matter bacterium in the *Quality Criteria for Water 1986* report, also known as the gold book. Based on the report, the geometric mean and single sample assay for *Enterococcus* in marine waters is not to exceed 35 per 100 mL and 104 per 100 mL, respectively (EPA, 1986). For freshwaters, the geometric mean and single sample assay for *E. coli* is not to exceed 126 and 235 per 100 mL, respectively, and *Enterococcus* is not to exceed 33 and 61 per 100 mL, respectively (EPA, 1986).

Limitations with Current Water Quality Criteria

Laws noted that there are several problems with the water quality criteria and sampling. First, the organisms that are being sampled are indicator bacteria and are not responsible for causing the health problems. Scientists rely on the fecal indicator bacteria because these are more abundant in feces than the actual pathogens and require smaller samples to detect their presence. Detection of the bacteria, viruses, protozoa, and worms that lead to gastrointestinal illnesses is much more difficult. Laws explained that 70 percent of hospitalizations from gastroenteritis in the United States remain due to unknown causes, especially because it can be quite expensive to assay for all the possible pathogens.

A second limitation is that traditional sampling tests take approximately 24 hours to obtain results because the cultures require approximately 24 hours to grow on the specified medium. For example, Boehm (2007) examined the correlation between one day's samples and the next day's samples at recreational beaches and found no correlation between the *Enterococcus* counts of consecutive samples. Laws noted that because of the time delays in testing, the results can indicate if it is safe to go in the water yesterday but give no indication of whether the water is safe today. In another study from Boehm and Weisberg (2005), the authors found that because of the 24-hour delay in assessing *Enterococcus* counts that beaches were often closed when the counts were acceptable. The closures followed days where the beaches were open and *Enterococcus* counts exceed 104 per 100 mL (Boehm and Weisberg, 2005). Additionally, Boehm (2007) has shown that there is extreme variability even between samples obtained every minute during 1-hour intervals, which raises questions about the adequacy of these water quality sampling tests.

Third, Laws noted that the assumption is that the fecal indicator bacteria come from human feces and at a specific time, but in reality there are multiple potential sources in the natural environment, including other mammals, birds, fish, and runoff from beach sands, soils, and plants (Muller et al., 2001; Yamahara et al., 2007). Some of these environmental reservoirs are not associated with feces or human pathogens at all, as seen with soils in Hawaii that have natural colonies of *Enterococcus* (Oshiro and Fujioka, 1995). In another study (Johnson et al., 2012) the authors examined two types of *Vibrio* bacteria (*V. parahaemolyticus* and *V. vulnificus*) found in recreational waters and which can cause gastrointestinal problems. The authors looked at concentrations of these bacteria in oysters and correlated this with the temperature and salinity of the oysters. Laws explained that 70 percent of the variance in the data could be explained based on the temperature and salinity of the oysters, which resulted in a strong correlation between the observed and predicted concentrations of the *Vibrio* bacteria in the oysters.

Failures of the Recreational Waters Monitoring System

Laws included two examples in his presentation, which describe important factors where the system failed when monitoring recreational waters in the United States and other countries. The first example is from Delhi, India, in 1955–1956. The public water supply became contaminated with sewage during the monsoon floods (Laws, 2000). The authorities ramped up the filtration and chlorination and assayed the treated water by looking for fecal indicator organisms. In this case, the water met the criterion for *E. coli* under indicated standards but approximately 20,000 clinical cases of hepatitis A were reported. Laws highlighted that viruses in general are more resistant to chlorine than bacteria, and in this example, infectious hepatitis or hepatitis A virus was particularly resistant. Because the treatment was targeted for *E. coli*, which is highly susceptible to chlorine, the authorities did not consider assessing for other pathogens that might have been present.

In 1993 in Milwaukee, approximately 400,000 people became ill after *Cryptosporidium parvum* (*C. parvum*) contaminated the public water supply. *C. parvum* is a protozoan that is a common waterborne pathogen. Chlorine is not an effective treatment mechanism because the *C. parvum* forms eggs that are resistant to chlorine (Corso et al., 2003), and water filtration is the only effective solution. Laws noted that this is another example where reliance on fecal indicator bacteria can be misleading about the water quality, and many people can become sick.

Laws explained that the EPA is working on two major factors to improve monitoring of recreational waters. The first involves finding a way to reduce the time between collecting the sample and obtaining the results. As noted above, the 24-hour assays have many limitations and do not provide all the necessary information. Real-time quantitative polymerase chain reaction (qPCR) has proven effective by detecting the DNA from the organism. However, the presence of DNA does not mean that the organism is present, which increases the possibility for false positives. For the EPA to consider utilizing this new assay, the determination must be supported by epidemiological studies that relate the assay results to human health risks.

Bacteriodes are the second factor in which the EPA is focusing its efforts. Bacteriodes are human-specific anaerobic bacteria and are a large component of human feces (they are present at higher concentrations than the coliforms that are currently used in sampling). Because they tend to be host specific, they can provide a clear target for DNA analysis and clear indication that the water was contaminated with human feces. Whereas they do not survive long in the environment, the viruses (bacteriophages) that infect the bacteriodes do persist for longer periods (e.g., weeks) and the assays are actually for the viruses, not the bacteriodes. Laws stated that this is the kind of indicator that is needed because it is present at constant levels through a period of time in the environment, it can survive sewage treatment, and it is a better indication of the presence of human enteric viruses that cause gastroenteritis.

In closing, Laws provided a summary of the problems identified in his presentation surrounding the monitoring of recreational waters that could be the focus of future efforts:

- address temporal and spatial variability (possibly develop new models),
- resolve nonspecificity of indicators (bacteroides are promising),
- find assays that mimic survival of the hardiest pathogens (bacteriophages may provide a better assay),
- need rapid results (qPCR is promising),
- address the fact that some human pathogens are unrelated to human feces, and
- relate new indicators to human health risks (requires epidemiological studies, which are expensive).

DISCUSSION

Bernie Goldstein from the University of Pittsburgh asked Laws to comment on the implications of the variability in beach monitoring data, noting that the third highest count of *Enterococcus* occurred on a day the beach was closed because it followed a day with a count exceeding the 104 per 100 mL limit. He continued to ask if the 104 count per 100 mL standard provides any indication of whether the bacteria level on the following day will exceed the established background level. Laws noted that typically these beach environments have a longshore current system so that what was present today will be downcurrent tomorrow, unless there is a continuous point source that is leaking into the water. So the nature of the input would determine the water quality on the following day for the recreational water.

REFERENCES

Boehm, A. B. 2007. *Enterococci* concentrations in diverse coastal environments exhibit extreme variability. *Environmental Science & Technology* 41(24):8227–8232.

Boehm, A. B., and S. B. Weisberg. 2005. Tidal forcing of *enterococci* at marine recreational beaches at fortnightly and semidiurnal frequencies. *Environmental Science & Technology* 39(15):5575–5583.

Cabelli, V. J. 1983. *Health effects criteria for marine recreational waters.* EPA-600/1-80-031. Research Triangle Park, NC: U.S. Environmental Protection Agency.

Corso, P. S., M. H. Kramer, K. A. Blair, D. G. Addiss, J. P. Davis, and A. C. Haddix. 2003. Cost of illness in the 1993 waterborne *Cryptosporidium* outbreak, Milwaukee, Wisconsin. *Emerging Infectious Disease* 9(4):426–431.

Costanza, R., R. d'Arge, R. de Groot, S. Farber, M. Grasso, B. Hannon, K. Limburg, S. Naeem, R. V. O'Neill, J. Paruelo, R. G. Raskin, P. Sutton, and M. van den Belt. 1997. The value of the world's ecosystem services and natural capital. *Nature* 387(6630):253–260.

EPA (U.S. Environmental Protection Agency). 1986. *Quality criteria for water 1986*. Washington, DC: Office of Water Regulations and Standards, EPA.

Johnson, C. N., J. C. Bowers, K. J. Griffitt, V. Molina, R. W. Clostio, S. F. Pei, E. Laws, R. N. Paranjpye, M. S. Strom, A. Chen, N. A. Hasan, A. Huq, N. F. Noriea, D. J. Grimes, and R. R. Colwell. 2012. Ecology of *Vibrio parahaemolyticus* and *Vibrio vulnificus* in the coastal and estuarine waters of Louisiana, Maryland, Mississippi, and Washington (United States). *Applied Environmental Microbiology* 78(20):7249–7257.

Laws, E. A. 2000. *Aquatic pollution: An introductory text.* New York: Wiley.

Muller, T., A. Ulrich, E. M. Ott, and M. Muller. 2001. Identification of plant-associated *enterococci*. *Journal of Applied Microbiology* 91(2):268–278.

Oshiro, R., and R. Fujioka. 1995. Sand, soil, and pigeon droppings: Sources of indicator bacteria in the waters of Hanauma Bay, Oahu, Hawaii. *Water Science and Technology* 31(5–6):251–254.

Yamahara, K. M., B. A. Layton, A. E. Santoro, and A. B. Boehm. 2007. Beach sands along the California coast are diffuse sources of fecal bacteria to coastal waters. *Environmental Science & Technology* 41(13):4515–4521.

8

New Approaches to Protect Ecosystem Services and Human Health

This chapter summarizes three presentations. The first presentation provides a brief overview of the *National Ocean Policy* and its coastal and marine spatial planning priority area which provide a framework for improving the management of often competing uses of natural water ecosystems. The second presentation provides a site-specific (Tampa Bay, Florida) example of ecosystem management around nutrient reduction and the improvement seen in the ecosystem and in ecosystem services. The third presentation examines the link between human population growth, human consumption, and potential strategies to optimize ecosystem services. The presentations are followed by a summary of the discussion that took place.

FRAMING OF MANAGEMENT OF ECOSYSTEM SERVICES

Paul Sandifer, Ph.D.
Chief Science Advisor, National Ocean Service,
National Oceanic and Atmospheric Administration

Dr. Paul Sandifer began the discussion by pointing out that our ocean, coasts, and the Great Lakes affect us all in a variety of ways. They support tens of millions of jobs and contribute trillions of dollars to the national economy each year (IOPTF, 2010; Kildow et al., 2009). They host a growing number of important activities, including recreation, science, commerce, transportation, energy development, and national security. They provide a wealth of natural resources and ecological benefits and services. These benefits and services also help to protect coastal communities from damaging floods and storms. Coastal wetlands, for example, shelter recreational and commercial fish species and serve as a natural filter to help keep waters clean. He emphasized that while a large part of the economy, health, and national security depends on oceans, coasts, and the Great Lakes, these water bodies face a wide range of threats and pressures. Overfishing, pollution, coastal development, and the impacts of climate change, for example, are placing more stress on wildlife and natural resources, as well as on people and coastal communities.

The importance of the oceans, coasts, and the Great Lakes and the need to safeguard them and their benefits and services is the focus of a number of federal policies. The July 2010 Executive Order 13547, Stewardship of the Ocean, Our Coasts, and the Great Lakes[1] and the final recommendations of the Interagency Ocean Policy Task Force (IOPTF) (2010), which together constitute the National Ocean Policy was a landmark step toward addressing the challenges that face the health of our ocean and coasts, and the Great Lakes and the economies they support. The National Ocean Policy is the culmination of decades of work that builds upon prior bipartisan efforts to address the many challenges facing the ocean, Sandifer said. These efforts recognized that there is a need to have better and more efficient ways to manage and tend to our ocean, coasts, and Great Lakes resources.

To improve federal coordination and leadership, the National Ocean Policy establishes a National Ocean Council comprising heads of executive departments and federal agencies. The council is charged to provide sustained high-level engagement to oversee implementation of the National Ocean Policy. It also has as a goal to improve working relationships among federal agencies across the spectrum (resource management, economic development, national security) as embodied by the panel. In addition, the National Ocean Policy emphasizes the importance of ongoing coordination at the federal, state, tribal, and local levels through a Governance Coordinating Committee and with stakeholders through the Ocean Research Advisory Panel.

The National Ocean Policy is focused around nine priority objectives that provide a bridge between the National Ocean Policy and action on the ground and in the water. The National Ocean Policy's nine priority objectives help focus limited federal resources on meeting the essential

---

[1] Executive Order 13547, Stewardship of the Ocean, Our Coasts, and the Great Lakes. 2010. Available at: http://www.whitehouse.gov/the-press-office/executive-order-stewardship-ocean-our-coasts-and-great-lakes (accessed July 15, 2013).

needs of Americans and ensuring demonstrable outcomes and results. Four of the nine priority objectives address overarching ways the U.S. government must operate differently to better improve stewardship. The second category of objectives is considered areas of special emphasis. The priority areas are listed in Table 8-1.

Sandifer pointed out that the priority area of coastal and marine spatial planning (CMSP) includes consideration of ocean and coastal ecosystem services. The IOPTF's Final Recommendations (2010) noted that CMSP is simply a planning framework to improve management of the myriad and often competing uses of ocean, coastal, and Great Lakes waters and resources. CMSP is to be science based, transparent, and informed by stakeholders and the public and to result in substantial economic, ecological, and social benefits. The Final Recommendations state that

> CMSP is intended to improve ecosystem health and services by planning human uses in concert with the conservation of important ecological areas, such as areas of high productivity and biological diversity; areas and key species that are critical to ecosystem function and resiliency; areas of spawning, breeding, and feeding; areas of rare or functionally vulnerable marine resources; and migratory corridors. Enhanced ecosystem services and benefits can be attained through CMSP because they are centrally incorporated into the CMS Plan as desired outcomes of the process and not just evaluated in the context of the individual Federal or State agency action. CMSP allows for a comprehensive look at multiple sector demands which would provide a more complete evaluation of cumulative effects. This ultimately is intended to result in protection of areas that are essential for the resiliency and maintenance of healthy ecosystem services and biological diversity, and to maximize the ability of marine resources to continue to support a wide variety of human uses.[2]

---

[2] Executive Order 13547, Stewardship of the Ocean, Our Coasts, and the Great Lakes. 2010. Available at: http://www.whitehouse.gov/the-press-office/executive-order-stewardship-ocean-our-coasts-and-great-lakes (accessed July 15, 2013).

TABLE 8-1 National Ocean Policy: Nine Priority Areas

How We Do Business	Areas of Special Emphasis
Ecosystem-based management	Resiliency and adaptation to climate change and ocean acidification
Coastal and marine spatial planning	Regional ecosystem protection and restoration
Informed decisions and improved understanding	Water quality and sustainable practices on land
Coordinate and support	Ocean, coastal, and Great Lakes observation, mapping, and infrastructure
	Changing conditions in the Arctic

Sandifer ended his presentation by noting that the National Ocean Policy specifically recognized that maintaining the health of our oceans, coasts, and the Great Lakes is essential for sustaining human health and well-being.

PROTECTING WATER QUALITY: TAMPA BAY, FLORIDA

Holly Greening, M.S.
Executive Director,
Tampa Bay Estuary Program

Holly Greening began by noting that her presentation would be a site-specific application of some of the new information coming out about ecosystem services and how that is being incorporated into decision making in Tampa Bay, Florida. Looking at ecosystem services and habitat restoration, Greening noted that habitat restoration is mainly done for fish and birds, but recovery of coastal habitats is also important for providing human services such as carbon sequestration, nutrient reduction, air quality improvements, and aesthetics. She stated that her presentation would focus specifically on the value provided to the Tampa Bay community from nutrient reduction as a result of coastal habitat recovery and restoration over the past 20 years.

Greening provided a brief overview of Tampa Bay (see Box 8-2). She explained that Tampa Bay has seen a dramatic recovery of the water quality, underwater seagrasses, and coastal habitats over the past 20 to 25

BOX 8-2
Facts About Tampa Bay

It is Florida's largest open water estuary.

The open water is about 400 square miles and the watershed is about 2,600 square miles.

It is a shallow estuary with an average water depth of 12 feet.

The watershed population currently is about 2.3 million.

The population has doubled in the past 20 years.

The port of Tampa is 1 of the 10 largest in the United States.

SOURCE: TBEP, 2006.

years. This is a real turnaround that many places are not experiencing right now, she said. The amount of nitrogen removed by the restored habitat over the past 20 years in Tampa Bay is equivalent to about two medium-size advanced wastewater treatment plants in terms of the amount of nutrient reduction provided by the habitat restoration and in avoided costs.

In the 1970s, Greening stated, Tampa Bay was heavily polluted with an excess of nitrogen, which resulted in vast amounts of macroalgae covering the estuarine surface water. These problems were highlighted in a *60 Minutes* exposé on nutrient pollution in the United States in the late 1970s. Half of Tampa Bay seagrasses were lost between 1950 and 1982 due to this excess nitrogen. The populations of macroalgae and phytoplankton covering the surface of the water blocked the sunlight and did not allow enough sunlight to reach the bottom where seagrasses were growing. Half of Tampa Bay's natural shoreline was altered during this period and 40 percent of the tidal marshes were destroyed. Some of the animal populations were negatively impacted as well; white ibis populations were reduced by about 70 percent, and fish kills were common.

Greening explained that most of the economic and environmental models from the 1970s indicated that no action could be taken to effectively recover the Tampa Bay estuary. However, the citizens were determined to do something and demanded action. In 1978, the Florida legislature required upgrades to all wastewater treatment plants in the Tampa Bay watershed (not just along the shoreline).[3] The wastewater

[3] Grizzle-Figg Act, Section 403.086, Florida Statutes.

treatment standards were also advanced and upgraded, which meant reducing the amount of nitrogen in the effluent to 3 milligrams per liter of total nitrogen. Typically, there were 20–30 milligrams per liter of total nitrogen present in these wastewater treatment plants before the revised standards. As a result of this Tampa also upgraded its sewage treatment plant. Greening noted that another option was to change to 100 percent water reuse, using effluent as irrigation water in the watershed, which was only implemented in St. Petersburg.

Greening emphasized that a key factor in this process was timing. It happened shortly after the Clean Water Act was passed and funds were available to Tampa to upgrade its wastewater treatment and to St. Petersburg to install the distribution lines. So in this instance, it was not just the local citizens' action, but also the availability of funds from the Clean Water Act that allowed the wastewater treatment plants to upgrade over a 3-year period.

Tampa Bay Estuary Program

In 1990, the Tampa Bay Estuary Program[4] was implemented and established as an intergovernmental program that uses measurable goals in science-based decisions. The policy board is made up of elected officials from the three counties in three major cities around Tampa Bay, as well the three regulatory agencies. Greening noted that one of the first things the policy board was asked to address was, "What do the citizens want from Tampa Bay? What do they want Tampa Bay to be?" One of the goals that came from a citizen survey indicated that the citizens wanted Tampa Bay to look more like it did in 1950 than it did in 1980. Greening explained that in 1950 the population was approximately one-fourth of the size as it was in 1980. Tampa Bay experienced a huge increase in population when air conditioning became available around 1950, but before then, the population growth rate was fairly flat.

In focusing on ecosystem services, the citizens identified three goals for restoring the estuary: (1) clear water, like the "good old days;" (2) better fishing; and (3) swimming without "seaweed" (macroalgae in this case). Greening noted that all three of these ecosystem services pointed to recovering underwater seagrasses by improving water quality and water clarity. The second major action taken by the policy board became

---

[4] Information on the Tampa Bay Estuary Program is available at: http://www.tbep.org (accessed September 9, 2013).

a very aggressive seagrass restoration goal to restore seagrass acres to that observed in 1950. Greening again highlighted that about half of the seagrasses were lost between 1950 and 1990 due to poor water quality. The map in Figure 8-1 shows the areas where seagrasses were present in 1950 but had disappeared by 1990. Most of the losses were from the deep edge, which indicates that the light was limited and the seagrasses were retreating to the shoreline.

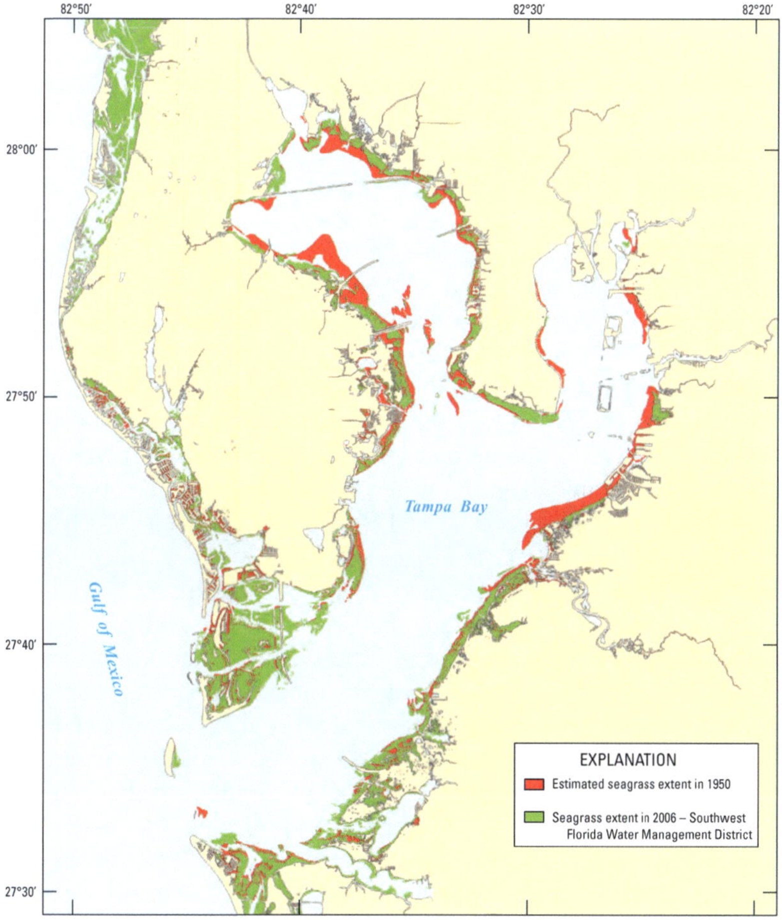

FIGURE 8-1 Estimated changes in seagrass cover in Tampa Bay between 1950 and 2006.
NOTE: Green areas show seagrass extent as mapped in 2006; red areas represent locations where seagrass occurred in 1950 but was not mapped in 2006.
SOURCE: Yates et al., 2011. Data from Haddad, 1989.

Greening explained that the policy board established two clear steps to accomplish the seagrass restoration goal: (1) identify light requirements for seagrass and (2) identify light attenuation factors. They found that the seagrass needs to receive about 22 percent of the light that hits the surface of the water in order to be able to thrive in Tampa Bay. This amount of light is vital for the seagrasses to survive and reproduce. They also found that chlorophyll, in the phytoplankton in the water column itself, was one of the major factors associated with reducing light. Turbidity and color, but not to the same extent as chlorophyll, also had an influence on light attenuation.

Greening emphasized the goal is to maintain 22 percent light to depth were the seagrasses are present and a light attenuation of chlorophyll concentration that will allow that light to be at depth. A series of mechanistic and empirical models were developed to relate to the chlorophyll target to a specific total nitrogen load. There is a chlorophyll target for each of the four major Bay segments and a total nitrogen load associated with maintaining and not exceeding that chlorophyll target.

During the decision process with the policy board, it was made clear that any impact on phytoplankton would affect the fish population that feeds on the phytoplankton and consequently would affect the bird population, more specifically, pelicans. The challenge was to balance the two goals surrounding the amount of light for seagrass growth and the amount of phytoplankton in the water column that would support many of the fish and bird species that feed on the fish and respectively on the phytoplankton. Greening stated that at the moment, a balance has been achieved, and both the seagrass and bird populations are recovering.

Importantly, Greening noted, the policy board realized that they could not control nitrogen alone in Tampa Bay. They needed to have local governments, regulatory agencies, local phosphate companies, agricultural interests, electric utilities, and all sources of nitrogen coming to the table and agreeing to meet the long-term goals of maintaining nitrogen loads that will allow for seagrass restoration. In the mid-1990s the Tampa Bay Nitrogen Management Consortium (TBNMC)[5] was formed, which is a public–private partnership with 40 participants that have accepted the responsibility for collectively meeting nitrogen load reduction goals. In just the past couple of years, TBNMC has also been meeting the U.S. Environmental Protection Agency (EPA) Total Maximum

[5] Information on the Tampa Bay Nitrogen Management Consortium is available at: http://www.tbeptech.org (accessed September 9, 2013).

Daily Load (TMDL) goals, which are the total amount of pollutants that a body of water can receive in 1 day. The consortium has developed and agreed on voluntary caps on all nitrogen loads from all sources. Those caps have now been incorporated into individual permits, which means that, even with increased population growth, local counties and private entities in Tampa Bay will continue to maintain nitrogen loads at that observed in 2003. The consortium has implemented many different projects in order to meet its long-term goals (about 250 in total), some of which have focused on decreasing industrial discharges, upgrading sewage plants, improving the air quality at power plants, improving the handling of toxic materials, improving stormwater treatment, and facilitating residential action. One interesting project, stated Greening, is where local entities and cities have implemented a residential fertilizer ban on nitrogen fertilizer during the rainy months in the summertime and also a point-of-sale restriction during that time, which have been successful.

As a result of all these different projects, Greening noted that there has been about a 50 percent reduction in the nitrogen loading between the 1970s and the 1990s in Tampa Bay, as shown on Figure 8-2. The distribution of the nitrogen sources has also changed over time. In the 1970s more than 50 percent of the nitrogen was attributed to point sources, most of which came from wastewater treatment plants. However, more difficult nonpoint sources are now the primary sources of nitrogen and are being targeted in Tampa Bay (e.g., the fertilizer ordinances that are in place).

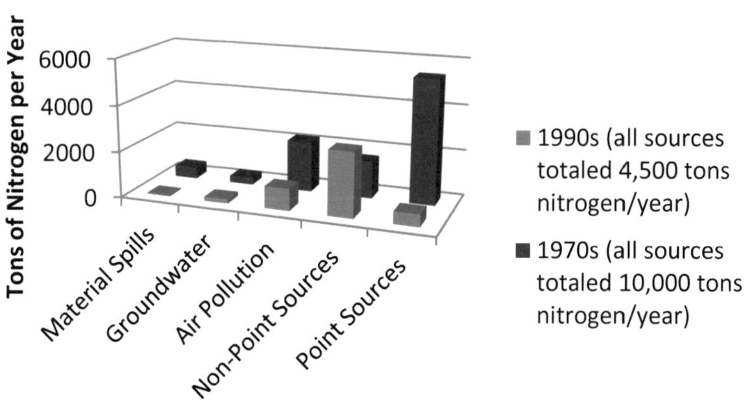

FIGURE 8-2 Nitrogen loading in the 1990s and 1970s in Tampa Bay.
SOURCE: Data from Greening and Janicki, 2006.

Water quality has also improved in the four major segments of Tampa Bay: Hillsborough Bay, Lower Tampa Bay, Middle Tampa Bay, and Old Tampa Bay. Starting in 1974, the data show that chlorophyll targets were not met. In 1984 some of the segments of the Bay began meeting the targets, but this was not consistent in all four segments until 1991. Greening stated that four major interventions contributed to this improvement:

- advanced wastewater treatment began in 1980,
- stormwater regulations were enacted in 1985,
- the Tampa Bay Estuary Program was established in 1990, and
- the Tampa Bay Nitrogen Management Consortium was initiated in 1995.

The seagrass recovery has also been progressing (see Figure 8-3). Greening explained that the goal identified by the policy board was to return to the level existing in 1950, which was approximately 40,000 acres of seagrass. The level in 1982 was about half of that observed in 1950. Slowly the seagrass has recovered; but, the biggest challenge was presented during the El Niño meteorological events of the late 1990s. During those years, Greening noted, the recovery goals were not achieved and a loss of approximately 2,000 acres of seagrass occurred. In the past few years, seagrass has recovered at a rate of about 500 acres per year and this is 500 acres of natural recovery not seagrass planting.

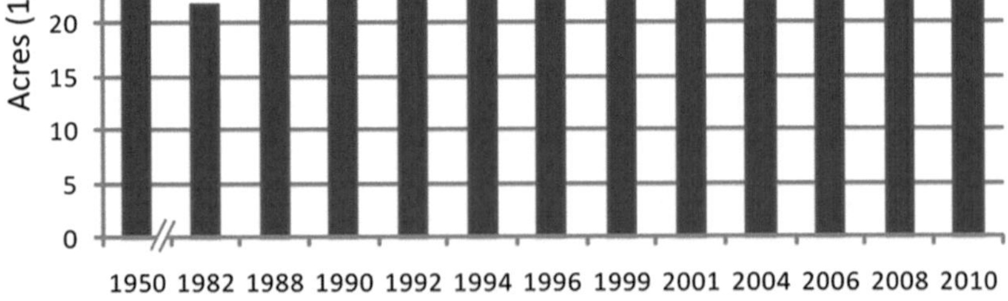

FIGURE 8-3 Tampa Bay seagrass recovery progress from 1982 through 2010. SOURCE: Greening, 2012. Data from Haddad, 1989.

Tampa Bay Ecosystem Services Pilot Project

Greening noted that Tampa Bay serves as a pilot project for the EPA's Ecosystem Services Research Program. One of the questions being asked is, "What is the value of the recovery in Tampa Bay?" Greening explained that not only have seagrasses recovered from the improved water quality, but salt marshes and mangroves have also recovered since 1990 primarily due to habitat restoration efforts. Cost assumptions have also been developed for the nutrient reduction piece of the ecosystem services in Tampa Bay. A conservative estimate shows that it costs $8.16 per pound of nitrogen for removal from traditional point sources (Birch et al., 2011) and the replacement cost for removing a pound of nitrogen from advanced wastewater treatment plants (as required in Tampa Bay) is $855 (Roeder, 2007). Additionally, the conservative valuation estimates for the seagrass, salt marsh, and mangrove restoration shows that about $5 million of value has been added compared to 1990 (Russell and Greening, 2013).

In closing, Greening stated that after 20 years of restoration and recovery in Tampa Bay, existing coastal habitats conservatively provide an estimated $24 million per year in nutrient reduction services for Tampa Bay watershed residents (this estimate may be up to 10 times higher if denitrification rates are actually closer to tropical seagrasses). In addition, she said, achieving habitat restoration and protection long-term goals in Tampa Bay could result in a conservative estimate of an additional $4 million per year in nutrient reduction services above current levels, making the total value $28 million per year. Again, this value is comparable to the cost of two wastewater treatment plants in the Bay.

Greening noted that ecosystem services may help with making difficult decisions about spending money on habitat restoration as untargeted improvements can be identified and valued (e.g., bacteriological improvements from nutrient reduction projects, and air quality improvements or carbon sequestration from mangrove restoration).

In summarizing the key elements in Tampa Bay's recovery, Greening emphasized that the first survey that indicated what the citizens wanted was critical in establishing science-based numeric goals and targets for Tampa Bay. Continued citizen involvement, regulation (e.g., Grizzle-Figg Act of 1990 and the EPA TMDL goals), public–private collaborative actions, long-term monitoring, valuation of ecosystem services, education and outreach, and assessment and adjustment have all been important to the adaptive management of ecosystem services in Tampa Bay.

OPTIMIZING ECOSYSTEM SERVICES IN THE FACE OF GLOBAL INCREASES IN HUMAN CONSUMPTION AND POPULATION GROWTH

G. David Tilman, Ph.D.
Regents Professor,
Department of Ecology, Evolution, and Behavior,
University of Minnesota

David Tilman stated that the focus of his presentation was to describe the environmental and coastal ecosystem impacts of a rapidly expanding human population. Within 50 years, Tilman said, the global population is projected to increase by 35 percent to about 9.5 billion inhabitants. At the same time, global per-capita incomes and gross domestic product (GDP) is projected to increase 140 percent; consequently, buying power and food consumption is also projected to increase. Associated with the increase in population, income, and consumption is the human domination of global ecosystems through the increase in food and energy demands, and environmental impacts of food and biofuels on coastal waterways (Tilman et al., 2011).

According to Tilman, rising per-capita incomes around the world drive the global demand for energy and contribute to fossil fuel greenhouse gas (GHG) releases. It is estimated that global fossil GHG emissions will rise from 7 gigatons per year to about 17 gigatons per year by 2050. In parsing out this relationship, Tilman explained that per-capita agricultural demand depends on per-capita income; as income rises, diets shift to a greater demand for crop calories or protein. Tilman described the results of a forecast study he conducted with colleagues that found a simple and temporally consistent relationship between per-capita GDP and per-capita demand for crop calories or protein among seven economic groups. The study found that per-capita use of calorie and protein for the richest nations were 256 percent[6] and

---

[6] Group B: Argentina, Chile, Greece, Israel, Italy, Malaysia, Mauritius, New Zealand, Portugal, Saudi Arabia, South Korea, Spain, Trinidad and Tobago, Uruguay, and Venezuela.

430 percent[7] greater than use by the poorest nations.[8] He noted that the difference between rich and poor countries is partly the result of greater meat consumption at higher incomes and low efficiency in the way some types of livestock convert crop calories and protein into edible food. Forecasts of future crop demand were developed based on the observed relationships between per-capita crop use and per-capita GDP. The analyses forecast a 110 percent increase for crop protein and a 100 percent increase for crop calories between 2005 and 2050 (Tilman et al., 2011). He added that there is a slight bias toward increasing the demand for protein in crops increasing more rapidly, such as from soybeans that are used as livestock feed and have a high level of protein.

Energy Use, Food Demand, and the Environment

Tilman emphasized that energy, food, and the environment are inextricably linked. For example, energy use is related to greenhouse emission and contributes to climate change which in turn affects storm intensity and crop growth. The increasing demand for food can increase the load of nitrogen and phosphorus fertilizers, and pesticides in coastal waters which are the home to major fisheries and sites for aquaculture. Turning to the agriculture sector, Tilman noted that producing the food needed to meet current needs releases about 32 percent of the total greenhouse gases released every year; the transportation sector contributes 14 percent. He emphasized that how demands for what is consumed are met will have a significant impact on future greenhouse gas emissions. Figure 8-4 shows the dramatic increase in total global use of nitrogen and phosphorus fertilizer, and area of global irrigated land during the Green Revolution (1960–2000). During that period, nitrogen use increased 600 percent, phosphorus 200 percent, and 80 percent more water was used in irrigation to double the global food supply. Similar trends are foreseen if the agriculture sector is expected to double food production in the next 50 years.

---

[7] Group A: Australia, Austria, Canada, Denmark, Finland, France, Germany, Ireland, Japan, The Netherlands, Norway, Sweden, Switzerland, United Kingdom, and United States.

[8] Poorest nations. Group F: Burkina Faso, Eritrea and Ethiopia, Gambia, Guinea, Haiti, Kenya, Madagascar, Malawi, Mali, Nepal, Rwanda and Burundi, Sudan, Tanzania, Togo, Uganda, Zambia, and Zimbabwe. Group G: Central African Republic, Chad, Democratic Republic of the Congo (former Zaire), Niger, and Sierra Leone.

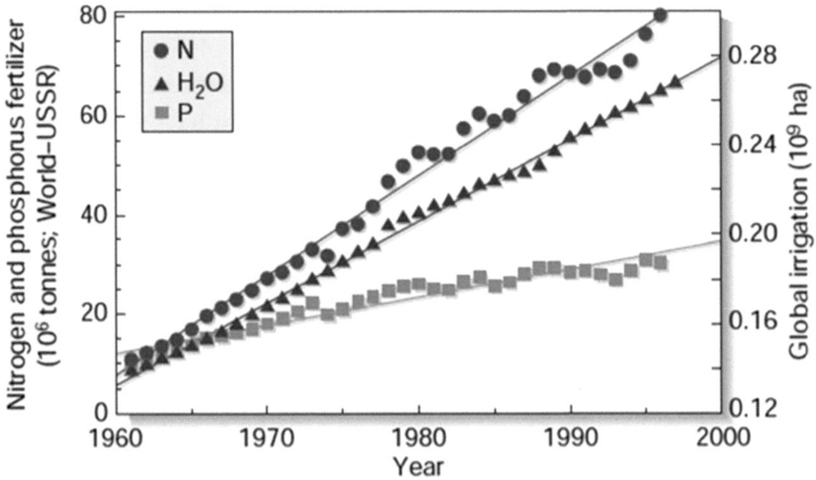

FIGURE 8-4 Use of nitrogen (N), phosphorus (P), and water (H_2O) to double the global food supply during the green revolution.
SOURCE: Tilman et al., 2002. Reprinted with permission from Nature Publishing Group.

There are important trade-offs related to increased food production, Tilman said. Agricultural practices to increase food production can have negative effects on the ecosystem. How crops are grown and which crops are grown are major determinants of the status of coastal waters and the services they provide to society. As noted earlier, fertilizers and pesticides can increase nutrients and toxins in groundwater and surface water. Between 1940 and 2000, pesticide production increased 950 percent (Tilman et al., 2002). Practices such as land clearing for agricultural purposes can degrade the soil quality, contribute to eutrophication of aquatic habitats, and reduce biodiversity. Between 1960 and 2000, 570 million hectares were cleared for agriculture (Tilman et al., 2001). According to Tilman, if global agriculture and diets trends remain similar to those of the past 40 years, it is estimated that by 2050 global nitrogen fertilizer use will increase by 185 percent, global phosphorus fertilizer use by 140 percent, global pesticide production by 170 percent, and an additional 800 million hectares of global land would be cleared for agricultural use (Tilman et al., 2001). Further, the impact of these practices would result in 185 percent more GHG emission; global agricultural GHG emissions in 2050 would almost equal total global fossil fuel emission now (Tilman et al., 2011).

Effects of Livestock Production on Ecosystems

Turning to the environmental effects of livestock production on agriculture and coastal waterways, Tilman said the trend in livestock production is toward high-density confinement production. With more and more animals housed in large-scale facilities, the handling and disposal of animal waste is a significant problem. Manure is often collected in lagoons that can release high levels of hydrogen sulfide and other toxic gases which can contaminate surface water and groundwater with nutrients, toxins, and pathogens.

Another important aspect of livestock production is the efficiency in animal protein production and the impact on agriculture. During the period 1960–2000, global per-capita meat production increased more than 60 percent (Tilman et al., 2002). Tilman further emphasized that there are significant differences in the amount of grain it takes to produce a kilogram (kg) of edible meat or the amount of plant protein needed to produce a kg of meat. In confined animal feeding operations, for cattle, 8 kg of grain are required for 1 kg of meat (8:1) and 20 kg of edible plant protein for 1 kg of meat protein (20:1). Poultry and many types of aquaculture fish are much more efficient sources of animal protein. Shifts in the types or amounts of meat produced and consumed can cause large shifts in grain demand, and the land and agrichemicals (nitrogen and phosphorus) needed to produce it, and large shifts would also occur in the potential impacts to the ecosystem, for example from nitrogen waste.

Changes in diet and food production can lead to important changes in nitrate loading. Nitrogen and phosphorus in animal feed is excreted as waste unless it is absorbed by the animal's body. The ratio of nitrogen (often as nitrate) in waste to nitrogen in edible protein varies considerably depending on the efficiency of the animal. The approximate ratio of nitrogen in waste to that in edible protein in cattle is about 19:1, for poultry it is about 3:1, and for some aquaculture fish it is about 2:1. Thus, per kilogram of edible protein, cattle production leads to much greater amounts of nitrogen excreted as waste than does fish or poultry production, Tilman said.

A number of countries have begun to develop strategies to increase nitrogen efficiency in the agriculture process such as applying the appropriate amount of fertilizer at the appropriate time during the growing season of the crop. For example, the European Union (EU) had water quality problems related to nitrogen loading in its coastal waters. They imposed laws requiring farmers to obtain permits to justify the purchase of nitrogen. As a result, there was a one-third decrease in amount of nitrogen being applied to the soil. Crop yields, however, continued to increase even though less nitrogen was applied; this is because nitrogen was applied more effectively, Tilman explained. More importantly, improvements in water quality measures were seen; nitrogen loading in the EU coastal waterways was significantly improved.

Tillman described a strategy to reduce the use of pesticides in agricultural production. In China, scientists discovered that they could eliminate the need to apply a toxic fungicide to rice by planting alternating rows of two different varieties of rice. One of the varieties was resistant to a major rice fungal disease. Its presence so greatly decreased disease incidence that fungicide use could be eliminated. This increased productivity and reduced human exposure to the pesticide.

In summarizing the presentation, Tilman reiterated that to meet human needs for food crop production there is an increased use of land, nitrogen and phosphorus fertilizers, and pesticides which affect coastal waterways and ultimately the oceans that also contribute to food production. The primary effects are related to nitrogen and phosphorus and GHGs that affect the atmosphere and the open oceans that provide fish and other seafood. This interrelated system is seen in Figure 8-5.

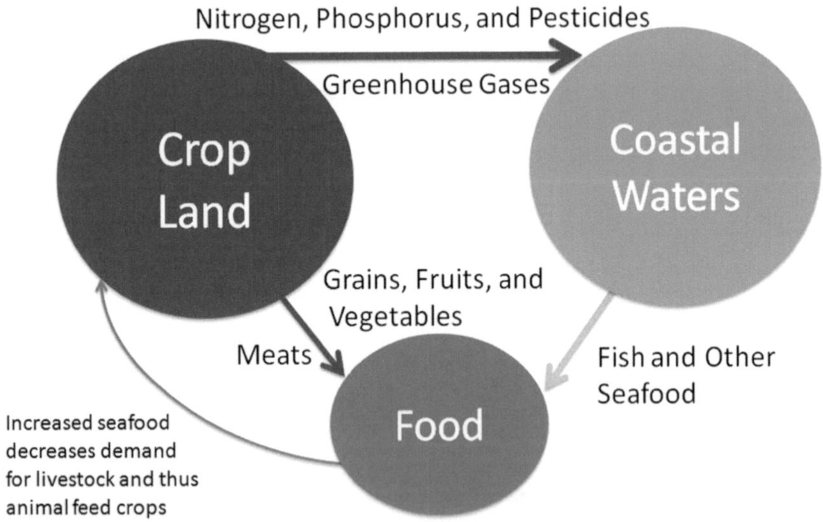

FIGURE 8-5 Food demand links to land and waters.
SOURCE: Tillman, 2012.

A system approach is needed to address the human demand for food and to lessen the impact of agriculture on the environment, Tilman said. One potential strategy is a dietary shift to higher-efficiency land animals, which would decrease nitrogen and GHG emissions. Another strategy is to increase fertilizer efficiency, which would also decrease nitrogen and GHG emissions. Implementing environmentally wise ways to better manage fisheries to increase the amount of fish and seafood to harvest and to include in the diet is another strategy. Moderate successes implementing all of these strategies would go far to meet the current and future human demand for food in ways that would help preserve the health of coastal waterways and oceans.

DISCUSSION

Lynn Goldman noted that the decision to meet the EPA TMDL goals in Tampa Bay presented by Greening is an effective way to address a problem like nitrogen, which has multiple sources that all impact one resource. She asked Greening about how population growth is being approached in Tampa Bay because growing communities will likely put out more nitrogen despite the fact that there have been improvements at the wastewater treatment plants and with fertilizer regulations. Greening explained that the Florida Department of Environmental Protection accepted the science behind the EPA TMDL goals for nitrogen management and allowed the public–private consortium to define how the goals would best be allocated throughout Tampa Bay. This resulted in much buy-in from the all the different partners representing point sources and nonpoint sources in the area. In terms of managing growth, Greening said, the Tampa Bay Estuary Program is looking at implementing projects that can offset new growth. In addition, she said, each of the permitted point and nonpoint sources are required to address this issue. If a permitted source needs to have an additional allocation beyond the 2003 cap, then it needs to show a nitrogen reduction somewhere else. Greening explained that some are offsetting the growth with additional stormwater treatment, fertilizer ordinances, and transfers of reduced emissions from the electric utilities in the watershed.

Richard Jackson from the University of California, Los Angeles, noted that the citizen involvement described in Tampa Bay supports the notion that people really want clean water. From a public health standpoint, he said, a public wastewater treatment system has tremendous benefits over independent septic systems because it is maintained and managed to recapture the wastewater. He asked Greening to comment. Greening stated that in Tampa Bay there are pockets of septic systems but that most of the residential areas are now on a central sewage system. One of the major sources of bacteriological contamination remains wastewater treatment package plants associated with trailer parks and older communities, she said, and those remain a target for both nitrogen and other public health reasons. Greening went on to say that in Tampa Bay approximately 70 to 80 percent of the wastewater treatment plants now have at least some recaptured water. One of the issues is that the piping to get this out to residential areas is quite expensive. As noted in her presentation, the reason that St. Petersburg was able change to 100 percent water reuse was a result of funding available from the Clean Water Act; without these funds this may not have been possible because of the associated infrastructure costs. Edward Laws from the Louisiana State University School of the Coast and Environment joined the discussion and noted that wastewater treatment plants are often not located with the idea of wastewater reuse in mind. This is now starting to change with resort areas investing in wastewater reuse for golf course irrigation but has not been a focus in other localities.

Laws then highlighted that the Chesapeake Bay Area has struggled to meet the goal for bay grass recovery and asked Greening to comment on why this program has been problematic. Greening stated that one main difference with Tampa Bay is that the majority of the population sees the bay nearly every day, which differs from the watershed placement in the Chesapeake involving multiple jurisdictions and populations that never see the bay. Because of this, the public support is much stronger in Tampa. She noted that another difference is scale; the Chesapeake Bay is much larger in size than Tampa Bay, which more closely resembles the size of Upper Chesapeake Bay. Greening explained that there have been successes in the Chesapeake Bay, but because of the scale issue they have not had the same impact as seen in Tampa. Additionally, the ecosystem services that benefit health and well-being have not been adequately valued in the Chesapeake Bay.

Laws asked Tilman to comment on the use of groundwater for irrigation, which is not sustainable in parts of China or the United States. Tilman noted that many aspects of global food production are not really sustainable and water is one of the big issues. For instance, about 11 percent of the land area in China is suitable for agriculture and they are utilizing about 13 percent at this time, including vast areas of dry land that require substantial amounts of groundwater that is not sustainable. In the United States, aquifers for corn production in the Midwest are not being recharged at the same rate as they are being drawn down, which is also a sustainability issue. There are many nations around the world that have low yields but good climate, but the farmers do not have the knowledge to do modern farming and may be subject to political unrest. Tilman stated he hopes these poorer nations of the world with large yield gaps can find a way to have the lands become more productive, which could potentially meet 70 percent of the food demand that is coming, without adding additional irrigation. So there are places in the world, he said, where we could have sustainable agriculture but these nations are lacking the knowledge, the capital, and the infrastructure that is required.

Dennis Devlin from ExxonMobil asked Tilman to comment on whether incorporating genetically modified organisms (GMOs) in his food production analysis would make a difference in the environmental impacts. Tilman noted that some GMOs provide a benefit if they can be grown without pesticides, but on the yield side, the GMOs are not providing greater yields compared to non-GMOs and in some instances provide lower yields, which can add costs. So currently, he said, incorporating GMO crops does not make much of a difference. Tilman went on to say that developing crops to remain ahead of diseases will continue to be a long-term issue and genetic modification will likely be an important tool as it has the potential to allow for the movement of many more genes to keep crops resistant to new pathogens.

REFERENCES

Birch, M. B. L., B. M. Gramig, W. R. Moomaw, O. C. Doering, III, and C. J. Reeling. 2011. Why metrics matter: Evaluating policy choices for reactive nitrogen in the Chesapeake Bay watershed. *Environmental Science and Technology* 45:168–174.

Greening, H. 2012. *Protecting water quality: Tampa Bay, Florida*. Presentation at the Institute of Medicine workshop on Understanding the Connections Between Coastal Waters and Ocean Ecosystem Services and Human Health: Basic Services, Valuation and Resiliency. Washington, DC.

Greening, H., and A. Janicki. 2006. Towards reversal of eutrophic conditions in a subtropical estuary: Water quality and seagrass response to nitrogen loading reductions in Tampa Bay, Florida, USA. *Environmental Management* 38(2):163–178.

Haddad, K. D. 1989. Habitat trends and fisheries in Tampa and Sarasota Bays. In *Tampa and Sarasota Bays: Issues, resources, and status and management*. NOAA Estuary of the Month Seminar Series 11 (pp. 113–138). Washington, DC: National Oceanic and Atmospheric Administration.

IOPTF (Interagency Ocean Policy Task Force). 2010. *Final recommendations of the Interagency Ocean Policy Task Force*. White House Council on Environmental Quality. Washington, DC: White House.

Kildow, T., C. S. Colgan, and J. Scorse. 2009. *State of the U.S. ocean and coastal economies*. National Ocean Economics Program. Available at: www.oceaneconomics.org (accessed August 14, 2013).

Roeder, E. 2007. *A range of cost-effective strategies for reducing nitrogen contributions from onsite sewage treatment and disposal systems*. Bureau of Onsite Sewage Programs. Tallahassee, FL: Florida Department of Health.

Russell, M., and H. Greening. 2013. Estimating benefits in a recovering estuary: Tampa Bay, Florida. *Estuaries and Coasts* doi: 10.1007/s12237-013-9662-8.

TBEP (Tampa Bay Estuary Program). 2006. *Charting the course: The comprehensive conservation and management plan for Tampa Bay*. St. Petersburg, FL: TBEP.

Tilman, D. 2012. *Optimizing ecosystem services in the face of global increases in human consumption and population growth*. Presentation at the Institute of Medicine workshop on Understanding the Connections Between Coastal Waters and Ocean Ecosystem Services and Human Health: Basic Services, Valuation and Resiliency. Washington, DC.

Tilman, D., J. Fargione, B. Wolff, C. D'Antonio, A. Dobson, R. Howarth, D. Schindler, W. H. Schlesinger, D. Simberloff, and D. Swackhamer. 2001. Forecasting agriculturally driven global environmental change. *Science* 292(5515):281–284.

Tilman, D., J. G. Cassman, P. A. Matson, R. Naylor, and S. Polasky. 2002. Agricultural sustainability and intensive production practices. *Nature* 418:671–677.

Tilman, D., C. Blazer, J. Hill, and B. L. Befort. 2011. Global food demand and the sustainable intensification of agriculture. *Proceedings of the National Academy of Sciences of the United States of America* 108(50):20260–20264.

Yates, K. K., H. Greening, and G. Morrison. 2011. Integrating science and resource management in Tampa Bay, Florida. *U.S. Geological Survey Circular* 1348:1–280.

9

Closing Observations

Paul A. Sandifer, Ph.D.
Chief Science Advisor, National Ocean Service,
National Oceanic and Atmospheric Administration

In closing, Paul Sandifer reminded the participants that this workshop began with an exploration of ocean ecosystem services, as well as how one begins to measure both ecosystem services and their connections to human health. He noted that the impacts of a variety of stressors were considered and there appears to be much more known about stressors' impacts on the environment and much less understood about the resulting health outcomes for humans. Both direct and indirect connections to human health are evident but measuring health outcomes remains more difficult.

Sandifer reminded the participants that during the workshop, great examples of site-specific issues were presented for Puget Sound, the Great Lakes, the Gulf of Mexico, and the Aral Sea. The talks highlighted challenges like storm protection and ecosystem restoration, as well as the potential for aquaculture, sustainable fisheries, and advances in other areas to contribute to improved global food security. Presentations effectively made connections among oceans, coast, land, land use, populations, and in some cases health. But again, he said, the metrics for human health impacts from diminished coastal ecosystem services are largely lacking. This should be a prime area for further exploration.

Sandifer noted that ecosystem services are the benefits to humans provided by ecosystems. He appreciated that this workshop looked at coastal ecosystem services in the specific context of how they support and affect human health and well-being, and considered how changes in ecosystems and related services can affect many benefits to humans that are taken for granted. Changes in ecosystems can have substantial effects on health and well-being, including mental health issues associated with job security or recreation or specific disease processes related to environmental exposures. Overall, this workshop made connections to a wide range of environmental health sciences topics and set the stage for further thinking about the nexus of ecosystem services and health, not only for this Roundtable but also with regard to the application of long-term research and investments in the environmental sciences.

A

Workshop Agenda

Understanding the Connections Between Coastal Waters and Ocean Ecosystem Services and Human Health: Basic Services, Valuation, and Resiliency

November 13–14, 2012
Lecture Room
National Academy of Sciences Building
2100 Constitution Avenue, NW
Washington, DC

November 13, 2012

8:30 a.m. **Welcome and Charge of the Workshop**

Lynn Goldman, M.D., M.P.H.
Vice Chair, Roundtable on Environmental Health Sciences, Research, and Medicine
Dean, George Washington University School of Public Health

Session 1: Overview of Ecosystem Services

Moderator: John Spengler, Ph.D., Akira Yamaguchi Professor of Environmental Health and Human Habitation, Department of Environmental Health, Harvard School of Public Health

8:45 a.m. **Understanding Ecosystem Services**

Lydia Olander, Ph.D.
Director, Ecosystem Services Program, Nicholas Institute for Environmental Policy Solutions
Adjunct Assistant Professor
Nicholas School of the Environment

9:10 a.m.	**Integration of Environmental Health and Marine Ecosystem Services in Puget Sound and Beyond** Tracy Collier, Ph.D. (via video link) Science Director Puget Sound Partnership
9:35 a.m.	**Discussion**
10:00 a.m.	**Break**
10:20 a.m.	**Stressors Impacting Coastal Waters and Ocean Ecosystem Services: Panel Discussion** Jonathan Garber, Ph.D. Acting Associate Director for Ecology National Health and Environmental Effects Research U.S. Environmental Protection Agency Paul Sandifer, Ph.D. Chief Science Advisor, National Ocean Service National Oceanic and Atmospheric Administration Ione Taylor, Ph.D. Associate Director, Energy and Minerals, and Environmental Health U.S. Geological Survey
11:20 a.m.	**Discussion**
11:50 a.m.	**Working Lunch**

Session 2: Opportunities for Medicine

Moderator: Lynn Goldman, M.D., M.P.H.

12:50 p.m.	**Current and Emerging Opportunities and Challenges for Pharmaceuticals from the Sea** William H. Gerwick, Ph.D. Professor, Skaggs School of Pharmacy and Pharmaceutical Sciences Scripps Institution of Oceanography
1:15 p.m.	**Discussion**

Session 3: Fishing, Fisheries, and Food

Moderator: Linda McCauley, Ph.D., R.N., Ph.D., FAAN, FAAOHN, Dean and Professor, School of Nursing, Emory University

1:40 p.m. **Marine Ecosystems Services and Fisheries**

Cynthia M. Jones, Ph.D.
A.D. and Annye L. Morgan Professor of Sciences
Eminent Scholar and Professor
Old Dominion University

2:05 p.m. **Seafood: Putting the World on a Fork**

Barton Seaver
Chef and National Geographic Fellow

2:30 p.m. **Aquaculture: Ensuring a Future Seafood Supply for a Healthy Population and Environment**

Kevan L. Main, Ph.D.
Senior Scientist and Director of Aquaculture Research
Mote Marine Laboratory
President, World Aquaculture Society

2:55 p.m. **Discussion**

3:40 p.m. **Break**

Session 4: Coastal Change and Human Health

Moderator: Jay Lemery, M.D., Assistant Professor, Attending Physician, and the Director of Wilderness and Environmental Medicine in the Department of Emergency Medicine, Weill Cornell Medical College

4:05 p.m. **The Role of Coastal Ecosystems in Protecting Gulf Coast Communities**

Denise J. Reed, Ph.D.
Chief Scientist
The Water Institute of the Gulf

4:30 p.m. **Discussion**

5:00 p.m. **Adjourn for the Day**

November 14, 2012

9:00 a.m. **Welcome Back**

Frank Loy, LL.B.
Chair, Roundtable on Environmental Health Sciences, Research, and Medicine

Session 5: Recreational Water and Water Quality

Moderator: Frank Loy, LL.B., Roundtable Chair

9:15 a.m. **Use of Indicator Organisms to Assess Public Health Risks Associated with Recreational Use of Natural Waters**

Edward Laws, Ph.D.
Professor
Louisiana State University School of the Coast and Environment

9:40 a.m. **Protecting Water Quality: Tampa Bay, Florida**

Holly Greening, M.S.
Executive Director
Tampa Bay Estuary Program

10:05 a.m. **Discussion**

10:35 a.m. **Break**

Session 6: New Approaches for Optimizing Sustainable Solutions in Ecosystem Services for Human Health

Moderator: Paul Sandifer, Ph.D.

10:50 a.m. **Framing of Management of Ecosystem Services**

Paul Sandifer, Ph.D.
Chief Science Advisor, National Ocean Service
National Oceanic and Atmospheric Administration

11:00 a.m. **New Opportunities for Resource Management: Life-Cycle Analysis, Sustainability, and Co-Benefits**

Steven A. Murawski, Ph.D.
Professor and Downtown Progress - Peter R. Betzer Endowed Chair of Biological Oceanography
University of South Florida

11:25 a.m.	**Applying a Systems and Sustainability Approach to Optimizing Ecosystem Services in the Face of Global Increases in Human Consumption and Population Growth** G. David Tilman, Ph.D. Regent Professor Department of Ecology, Evolution, and Behavior University of Minnesota
11:50 a.m.	**Discussion**
12:20 p.m.	**Summation** Paul Sandifer, Ph.D.
12:30 p.m.	**Adjourn**

B

Speaker Biographical Sketches

Tracy Collier, Ph.D., is the science director for the Puget Sound Partnership and a visiting scientist at the Center for Urban Waters in Tacoma. He also serves as the Science Advisor for the National Oceanic and Atmospheric Administration's (NOAA's) Oceans and Human Health Program. There, he provides science direction in the areas of chemical contaminants, pathogens, and algal toxins, and their effects on human and ecosystem health. Until 2010, he was an environmental toxicologist at the NOAA Northwest Fisheries Science Center. Dr. Collier received his Ph.D. from the University of Washington in 1988.

Jonathan Garber, Ph.D., is the acting associate director for ecology in the National Health and Environmental Effects Research Laboratory (NHEERL) at the U.S. Environmental Protection Agency (EPA). NHEERL is the EPA's focal point for scientific research on the effects of contaminants and environmental stressors on human health and ecosystem integrity. Its research mission helps the EPA to identify and understand the processes that affect our health and environment, and assists the EPA to evaluate the risks that pollution poses to humans and ecosystems.

William Gerwick, Ph.D., is a professor at the Skaggs School of Pharmacy and Pharmaceutical Sciences at the Scripps Institute of Oceanography. His research focuses on exploring the unique natural products of marine algae and cyanobacteria for useful biomedical properties in the areas of cancer, inflammation, neurochemical pathways, and infectious disease, including tropical diseases such as malaria and leishmaniasis. Previously, he was the deputy director of the Marine Freshwater Biomedical Center at Oregon State University. Dr. Gerwick received a Ph.D. in oceanography in 1981 from the Scripps Institution of Oceanography at the University of California, San Diego.

Lynn R. Goldman, M.D., M.P.H., is an American public health physician, trained as a pediatrician and epidemiologist. Now dean of the George Washington University School of Public Health, she is perhaps best known for her role in helping craft the Food Protection Act passed by Congress in 1996, the first national environmental law to explicitly require measures to protect children from pesticides. In 1993, Dr. Goldman was appointed by President Bill Clinton and confirmed by the U.S. Senate as assistant administrator for toxic substances at the EPA, becoming the first physician to serve in this capacity. During her 5 years at the EPA, from 1993 to 1998, she promoted pesticide legislation reform, assessment of industrial-chemical hazards, and children's health issues. Dr. Goldman is a graduate of the University of California, San Francisco, School of Medicine and Johns Hopkins Bloomberg School of Public Health.

Holly Greening, M.S., serves as executive director of the Tampa Bay Estuary Program (TBEP). She is responsible for maintaining the strong public and private partnerships forged through TBEP for the continuation of the bay's science-based restoration and recovery strategies. She served on the Governing Board of the Estuarine Research Federation, the National Academy of Sciences Ocean Studies Board, and three National Research Council Committees on coastal issues. She is currently serving on the Florida Oceans and Coasts Council and as associate editor for the journal *Estuaries and Coasts*. Ms. Greening has an M.S. in marine ecology from Florida State University.

Cynthia Jones, Ph.D., is the A.D. and Annye L. Morgan professor of sciences, professor of ocean, earth, and atmospheric sciences, an eminent scholar, and the director of the Center for Quantitative Fisheries Ecology at Old Dominion University. She studies marine fisheries and the quantitative ecology of fish. She is a fellow of the American Association for the Advancement of Science and has been recognized numerous times for faculty excellence. Dr. Jones received a Ph.D. in oceanography in 1984 from the University of Rhode Island.

Edward Laws, Ph.D., is a professor in the Department of Environmental Sciences at the Louisiana State University School of the Coast and Environment. He also serves as the director of the Pacific Research Center for Marine Biomedicine, a Center for Oceans and Human Health supported by the National Science Foundation and National Institute of

Environmental Health Sciences. Previously, he taught in the Department of Oceanography at the University of Hawaii for 30 years. He has published more than 140 papers in scholarly journals and is the author of a textbook on aquatic pollution that has been translated into numerous languages. Dr. Laws received his Ph.D. in chemical physics from Harvard University in 1972.

Jay Lemery, M.D., is an assistant professor of emergency medicine at the University of Colorado. He has an academic expertise in the field of wilderness and environmental medicine and was the director of Cornell Wilderness & Environmental Medicine from 2005–2012. He holds an appointment at the Harvard School of Public Health at the FXB Center for Health and Human Rights and serves a contributing editor for its journal *Health and Human Rights*. He is a consultant for the Climate and Health Program at the Centers for Disease Control and Prevention and became president of the Wilderness Medical Society in the summer of 2012. Dr. Lemery received his M.D. from Dartmouth Medical School.

Frank Loy, LL.B., has served in the Department of State in four administrations. His portfolio included developing U.S. international policy and conducting negotiations in the fields of the environment and climate change, human rights, the promotion of democracy, refugees and humanitarian affairs, and counter-narcotics. In 2011 President Obama named him the U.S. Alternate Representative to the U.N. General Assembly. At present he serves on the boards of numerous nonprofit organizations. In the field of the environment these include Resources for the Future (former chair), Environmental Defense Fund (former chair), The Nature Conservancy, C2ES, and ecoAmerica (chair). He also chairs the boards of Population Services International and the Arthur Burns Fellowship Program and serves on the boards of the American Institute for Contemporary German Studies and The Washington Ballet.

Kevan Main, Ph.D., is the manager for the Marine and Freshwater Aquaculture Research Program at Mote Marine Laboratory. She directs the operations at Mote's 200-acre field station, Mote Aquaculture Park, in eastern Sarasota County. In September 2012, she became president of the World Aquaculture Society. Her research interests are in developing sustainable aquaculture methods to produce marine fish, sturgeon, shrimp, abalone, and corals. She has more than 20 years of experience in the aquaculture of tropical and subtropical fish and invertebrates from around the world, and has published 7 books and authored more than 50 peer-reviewed publications. Dr. Main received her Ph.D. from Florida State University.

Linda McCauley, Ph.D., R.N., FAAN, FAAOHN, is dean and professor at Emory University's Nell Hodgson Woodruff School of Nursing, and professor in the Department of Environmental Health at the Rollins School of Public Health. She is a nationally recognized leader in nursing education and in research on environmental exposures and health hazards among vulnerable populations, including workers and young children. She is a fellow of the American Academy of Occupational Health Nurses and the American Academy of Nursing. Dr. McCauley received her master's degree in nursing from Emory University and her Ph.D. in environmental health from the University of Cincinnati.

Steven Murawski, Ph.D., is a professor and the Downtown Progress - Peter R. Betzer Endowed Chair of Biological Oceanography at the University of South Florida. He is a fisheries biologist and marine ecologist involved in understanding the impacts of human activities on the sustainability of ocean ecosystems. He has developed approaches for understanding the impacts of fishing on marine fish complexes exploited in mixed-species aggregations. He is also a U.S. delegate and currently a vice-president of the International Council for the Exploration of the Sea, a 20-nation organization dedicated to increasing understanding of ocean ecosystems. Dr. Murawski received his Ph.D. from the University of Massachusetts Amherst in 1984.

Lydia Olander, Ph.D., is the director of the Ecosystem Services Program at the Nicholas Institute for Environmental Policy Solutions. She is currently developing the Nicholas Institute and Duke University's expanding initiative on ecosystem services. She is coordinating Duke's Ecosystem Services Working Group, the development of a National Ecosystem Services Partnership, and the Institute's programs on greenhouse gas offsets. She directs the Technical Working Group on Agricultural Greenhouse Gases and, when time permits, works on the burgeoning multinational effort on reduced emissions from deforestation and degradation. Dr. Olander received a Ph.D. in biogeochemistry from Stanford University in 2002 and a master's in forest science from Yale University in 1995.

Denise Reed, Ph.D., is the interim director of the Pontchartrain Institute for Environmental Sciences and a professor in the Department of Earth and Environmental Sciences at the University of New Orleans. Her research focuses on various aspects of sediment dynamics in coastal wetlands, with emphasis on sediment mobilization and marsh hydrology, both natural and altered, as factors controlling sediment deposition. She has participated in numerous research projects concerning marsh and estuarine sediment dynamics on the Gulf and Pacific coasts of the United States as well as in Europe and South America. Dr. Reed received a Ph.D. in geography from the University of Cambridge, England.

Paul Sandifer, Ph.D., is the chief science advisor for NOAA's National Ocean Service. He is currently leading a NOAA-wide effort in ecological forecasting, development of a NOAA "Gulf Coast Ecosystem Restoration Science, Observation, Monitoring and Technology Program" under the RESTORE Act of 2012, a NOAA health strategy, and establishment of a science career track for NOAA employees, among other duties. His research interests include ocean policy, coastal ecosystem science, and aquaculture, and he is recognized as one of the architects of the "metadiscipline" of oceans and human health. He is currently co-chair of the interagency Task Force on Integrating Science and Technology for Sustainability. Prior to coming to NOAA, Dr. Sandifer had a distinguished 31-year career with the South Carolina Department of Natural Resources, including service as agency director. He is a member of graduate faculties at the College of Charleston and the Medical University of South Carolina and has an extensive publication record.

Barton Seaver is a chef and National Geographic Fellow who has dedicated his career to restoring the relationship we have with our ocean. It is his belief that the choices we are making for dinner are directly impacting the ocean and its fragile ecosystems. He was *Esquire* magazine's 2009 "Chef of the Year" and was honored as a "StarChefs.com Community Innovator," as voted by more than 1,000 chefs and culinary leaders worldwide. In his first book, *For Cod and Country*, he introduces an entirely new kind of casual cooking featuring seafood that has not been overfished or harvested using destructive methods.

John D. Spengler, Ph.D., is the Akira Yamaguchi Professor of Environmental Health and Human Habitation in the Department of Environmental Health at Harvard University's School of Public Health.

He is the director of the Sustainability and Environmental Management Program for Harvard Extension School and the Center for Health and the Global Environment at Harvard School of Public Health. He has conducted research in the areas of personal monitoring, air pollution health effects, aerosol characterization, and indoor air. More recently, Dr. Spengler has been involved in research that includes the integration of knowledge about indoor and outdoor air pollution as well as other risk factors into the design of housing, buildings, and communities. He received a B.S. in physics from the University of Notre Dame, an M.S. in environmental health sciences from Harvard University, and a Ph.D. in atmospheric sciences from the State University of New York–Albany.

Ione Taylor, Ph.D., serves as associate director for the U.S. Geological Survey (USGS) Energy and Minerals, and Environmental Health Mission Area with oversight for energy and mineral resource programs, toxics hydrology, and contaminants biology programs. She began her career working for the New Mexico Bureau of Mines on mineralization in volcanic systems and spent 15 years as a petroleum industry geologist in domestic and international hydrocarbon exploration, research and development, and senior leadership roles. She joined the USGS in 1999 as Science Center Director, Eastern Energy Resources, and she has served as Deputy Regional Director and Regional Chief Scientist for Eastern Region, and Chief Scientist for Geography. Dr. Taylor holds B.S., M.S., and Ph.D. degrees, an Executive Certificate from MIT, and Senior Executive Service certification from the Department of the Interior.

G. David Tilman, Ph.D., is one of the world's leading ecologists, blending theoretical and experimental work seamlessly. His classic research created the benchmark model for determining how different organisms within an ecosystem compete for resources. His field experiments and theoretical insights have helped to alert scientists to the fact that the reduction in the number of plant and animal species on the planet has a profound effect on the way Earth's ecosystems function. For the past decade, much of his work has focused on the environmental impacts of global agriculture and how global food demand could be met in more sustainable ways. Dr. Tilman is the Regent's Professor and McKnight Presidential Chair in Ecology at the University of Minnesota, as well as an instructor in conservation biology; ecology, evolution, and behavior; and microbial ecology. He is director of the Cedar Creek

Ecosystem Science Reserve Long-term Ecological Research Station. He earned his Ph.D. at the University of Michigan in 1976.